Bifurcation Dynamics of a Damped Parametric Pendulum

Synthesis Lectures on Mechanical Engineering

Synthesis Lectures on Mechanical Engineering series publishes 60–150 page publications pertaining to this diverse discipline of mechanical engineering. The series presents Lectures written for an audience of researchers, industry engineers, undergraduate and graduate students.
Additional Synthesis series will be developed covering key areas within mechanical engineering.

© Springer Nature Switzerland AG 2022

Reprint of original edition © Morgan & Claypool 2020

Bifurcation Dynamics of a Damped Parametric Pendulum

Yu Guo and Albert C.J. Luo

ISBN: 978-3-031-79644-9   paperback
ISBN: 978-3-031-79645-6   ebook
ISBN: 978-3-031-79646-3   hardcover

DOI 10.1007/978-3-031-79645-6

A Publication in the Springer series
SYNTHESIS LECTURES ON MECHANICAL ENGINEERING

Lecture #22
Series ISSN
Print 2573-3168   Electronic 2573-3176

Scan this code to view the animation, or view online at:
www.morganclaypool-publishers.com/Guo_Luo/

Bifurcation Dynamics of a Damped Parametric Pendulum

Yu Guo
Midwestern State University

Albert C.J. Luo
Southern Illinois University

SYNTHESIS LECTURES ON MECHANICAL ENGINEERING #22

ABSTRACT

The inherent complex dynamics of a parametrically excited pendulum is of great interest in nonlinear dynamics, which can help one better understand the complex world. Even though the parametrically excited pendulum is one of the simplest nonlinear systems, until now, complex motions in such a parametric pendulum cannot be achieved. In this book, the bifurcation dynamics of periodic motions to chaos in a damped, parametrically excited pendulum is discussed. Complete bifurcation trees of periodic motions to chaos in the parametrically excited pendulum include:

- period-1 motion (static equilibriums) to chaos, and

- period-m motions to chaos ($m = 1, 2, \cdots, 6, 8, \cdots, 12$).

The aforesaid bifurcation trees of periodic motions to chaos coexist in the same parameter ranges, which are very difficult to determine through traditional analysis. Harmonic frequency-amplitude characteristics of such bifurcation trees are also presented to show motion complexity and nonlinearity in such a parametrically excited pendulum system. The non-travelable and travelable periodic motions on the bifurcation trees are discovered. Through the bifurcation trees of travelable and non-travelable periodic motions, the travelable and non-travelable chaos in the parametrically excited pendulum can be achieved. Based on the traditional analysis, one cannot achieve the adequate solutions presented herein for periodic motions to chaos in the parametrically excited pendulum. The results in this book may cause one rethinking how to determine motion complexity in nonlinear dynamical systems.

KEYWORDS

parametric pendulum, bifurcation trees to chaos, periodic motions, frequency-amplitude characteristics, non-travelable periodic motion, travelable periodic motions

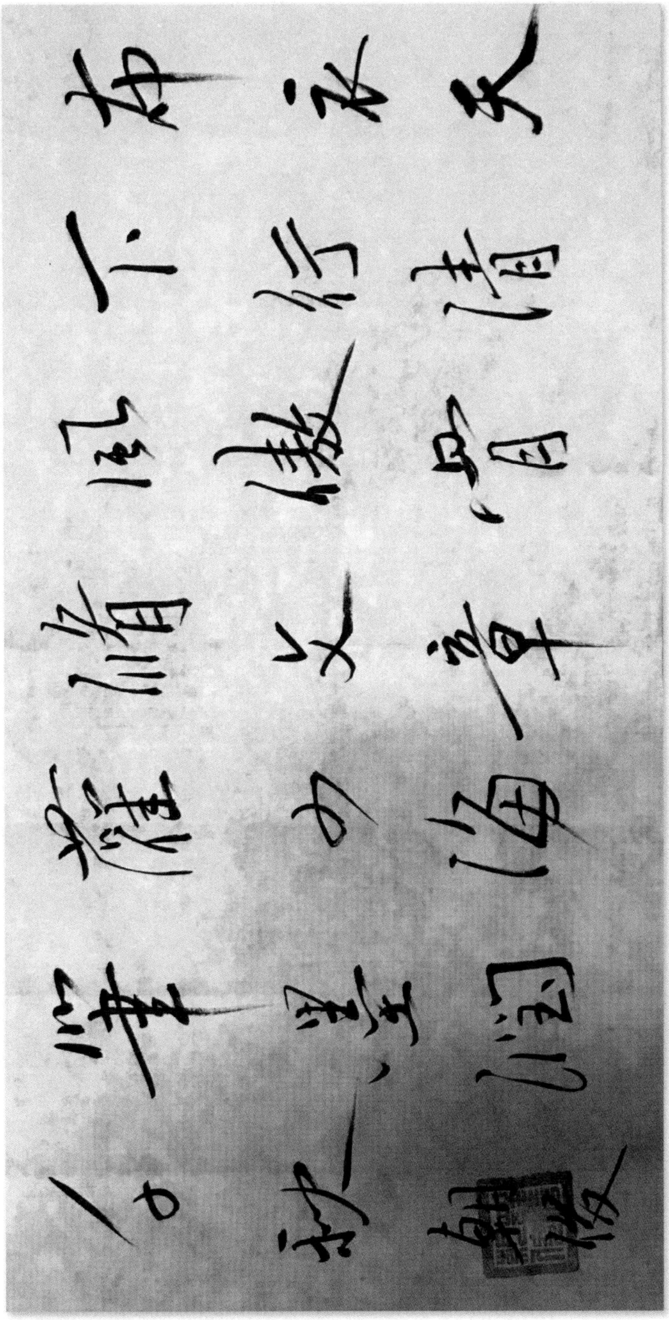

Contents

Preface

It is significant to determine periodic motions to chaos in a parametric pendulum even though the parametrically excited pendulum is one of the simplest nonlinear systems. This is because the inherent complex dynamics of the parametrically excited pendulum helps one better understand the complex world. However, until now, complex motions in the parametrical pendulum cannot be achieved yet through the traditional analysis. What are the mechanism and mathematics of such complex motions in the parametric pendulum? For polynomial nonlinear dynamical systems, one recently used the generalized harmonic balance method to obtain the analytical solutions of periodic motions. For non-polynomial, nonlinear dynamical systems, one still has difficulty to get the adequate solutions of periodic motions through the generalized harmonic balance method. Thus, one developed an implicit mapping method for periodic motions in nonlinear dynamical systems. Such an approach can be used to analytically predict periodic motions in the non-polynomial dynamical systems. Thus, the bifurcation dynamics of the parametric pendulum can be studied, and the corresponding results can help one understand motion complexity in nonlinear dynamical systems. In this short book, the bifurcation dynamics of periodic motions to chaos in a damped, parametrically excited pendulum is discussed through a semi-analytical method. The method is based on discretization of differential equations of the dynamical system to obtain implicit maps. Using the implicit maps, mapping structures are developed for specific periodic motions, and the corresponding periodic motions can be predicted analytically. The complete bifurcation trees of periodic motions to chaos varying with excitation frequency are presented.

This book has seven chapters. Chapter 1 provides an introduction. The implicit mapping method for periodic motions in nonlinear dynamical systems is presented in Chapter 2. Based on the semi-analytical method, periodic motions in the parametrically excited pendulum are discussed in Chapter 3. In Chapter 4, bifurcation trees of periodic motions to chaos are presented for period-1 static points, period-1, period-2, ..., period-6, period-8, period-10, and period-12 motions to chaos. The corresponding harmonic analysis of the bifurcation trees is presented in Chapter 5. Travelable and non-travelable periodic motions of the parametric pendulum are illustrated in Chapters 6 and 7, respectively. For non-travelable periodic motions, the corresponding harmonic amplitudes and phases of displacement are also be illustrated. For travelable periodic motions, the harmonic analysis of the travelable periodic motions is based on the velocity rather than displacement because the initial and ending displacements do not equal each other. This book will provide a clear skeleton of bifurcation trees of periodic motions to chaos in the parametric pendulums.

Finally, the authors hope the materials presented herein can last long for science and engineering. Some results presented in this book may give inspiration on the nonlinear dynamics community.

Yu Guo, Wichita Falls, Texas
Albert C.J. Luo, Edwardsville, Illinois
November 2019

CHAPTER 1

Introduction

It is significant to determine periodic motions to chaos in a parametric pendulum even though the parametrically excited pendulum is one of the simplest nonlinear systems. This is because the inherent complex dynamics of the parametrically excited pendulum helps one better understand the complex world. However, until now, complex motions in the parametrical pendulum cannot be achieved yet through the traditional analysis. What are the mechanism and mathematics of such complex motions in the parametric pendulum? For polynomial nonlinear dynamical systems, one recently used the generalized harmonic balance method to obtain the analytical solutions of periodic motions. For non-polynomial, nonlinear dynamical systems, one still has difficulty getting the adequate solutions of periodic motions. Herein, the implicit mapping method will be used to analytical predict the periodic motions in the non-polynomial dynamical systems. The parametric pendulum will be as an example to be investigated, and the corresponding methodology and results can help one understand motion complexity in nonlinear dynamical systems. A parametric pendulum system is very simple but it possesses rich and complicated dynamical behaviors, and such dynamical behaviors exist in other nonlinear oscillators.

The early studies on periodic motion in nonlinear systems were mainly based on the perturbation methods. In 1788, Lagrange [1] investigated the three-body problems as a perturbation of the two-body problems. In 1899, Poincare [2] developed the perturbation theory and applied such a method for the periodic motions of celestial bodies. In 1920, van der Pol [3] used the method of averaging for the periodic solutions of an oscillator circuit. In 1928, Fatou [4] proved the asymptotic validity of the method of averaging from the solution existence theorems of differential equations. 1935, Krylov and Bogoliubov [5] extended the method of averaging to nonlinear oscillations in nonlinear vibration systems. In 1964, Hayashi [6] presented the perturbation methods including averaging method and the principle of harmonic balance method for nonlinear oscillations. In 1969, Barkham and Soudack [7] used an extended Krylov–Bogoliubov method to obtain the approximated solutions of a second-order nonlinear autonomous differential equations. In 1987, Garcia–Margallo and Bejarano [8] determined the approximated solutions of nonlinear oscillations using a generalized harmonic balance method. Yuste and Bejarano [9, 10] used the elliptic functions instead of trigonometric functions to improve the Krylov–Bogoliubov method on nonlinear oscillators. In 1990, Coppola and Rand [11] used the averaging method with elliptic functions to obtain the approximation of limit cycles.

One has been interested in periodic motions and chaos in a parametrically excited pendulum system. In 1981, Leven and Koch [12] started to show chaotic behaviors in the parametric pendulum, and McLaughlin [13] numerically presented period doubling bifurcations in a parametrically excited pendulum. In 1985, Koch and Leven [14] used the perturbation method to study the boundaries of subharmonic and homoclinic bifurcations in a parametrically excited pendulum. Clifford and Bishop [15, 16] investigated travelable and non-travelable periodic motions in such a pendulum system. In 1998, Sanjuan [17] studied the subharmonic bifurcations of a parametrically excited pendulum with a non-harmonic perturbation. Luo [18, 19] investigated chaotic motions and the resonance stochastic layer of a parametrically excited pendulum. In 2003, Garira and Bishop [20] presented the solutions of travelable periodic motions in the parametrically excited pendulum. Such travelable orbits and the corresponding stability of the parametric pendulum can be also found in [21–23]. In 2007, Lu [24] studied the existence of chaos in a parametrically excited, undamped pendulum system using the shooting method.

On the other hand, researchers have been interested in analytical predictions of periodic motions for nonlinear oscillatory systems. In 2012, Luo [25] developed an analytical method for analytical solutions of periodic motions in nonlinear dynamical systems, which was based on the generalized harmonic balance. The detailed discussion can be found in Luo [26–28]. Luo and Huang [29] used the generalized harmonic balance method for the analytical solutions of period-1 motions in the Duffing oscillator with a twin-well potential. Luo and Huang [30] also employed a generalized harmonic balance method to find analytical solutions of period-m motions in such a Duffing oscillator. The analytical bifurcation trees of periodic motions in the Duffing oscillator to chaos were obtained (see also Luo and Huang [31–36]). Such analytical bifurcation trees showed the connection from periodic motions to chaos analytically. Luo and Yu [37] also employed the generalized harmonic balance method for approximate analytical solutions of period-1 motions in a nonlinear quadratic oscillator. Analytical solutions of periodic motions in van der Pol oscillator were also presented by Luo and Laken [38, 39]. The analytical solutions of period-m motion to chaos in the van der Pol–Duffing oscillator were studied in Luo and Laken [40]. In 2016, Luo and Yu [41, 42] discussed the analytical solutions for the bifurcation trees of period-1 motions to chaos in a two-degree-of-freedom nonlinear oscillator. In 2013, Luo [43] extended such ideas of the generalized harmonic method for periodic motions in time-delay, nonlinear dynamical systems. Luo and Jin [44] applied such a method for the bifurcation tree of period-1 motion to chaos in a periodically forced, quadratic nonlinear oscillator with time-delay. Further, Luo and Jin [45–47] investigated the periodic motions to chaos in a time-delayed Duffing oscillator. The generalized harmonic balance method is suitable for periodic motions in nonlinear systems with polynomial nonlinearity.

However, such a method is difficult to be applied to nonlinear dynamical systems with non-polynomial functions (i.e., pendulum system). Thus, in 2015, Luo [48] developed a semi-analytical method for periodic motions in complicated nonlinear dynamical systems. Luo and Guo [49, 50] applied such a method to predict the bifurcation trees of periodic motions in a

Duffing oscillator. Luo and Xing [51, 52] employed the method for symmetric and asymmetric period-1 motions and bifurcation trees of period-1 motions to chaos in a delayed, hardening Duffing oscillator. Luo and Xing [53] investigated the time-delay effects for periodic motions in a time-delay, Duffing oscillator. Furthermore, Xing and Luo [54, 55] subsequently studied the periodic motions in a twin-well, Duffing oscillator with time-delay displacement feedback and discussed the possibility of infinity bifurcation trees in such a nonlinear oscillator. To study higher-order dynamical systems, Xu and Luo [56] applied this method for a coupled van der Pol Duffing oscillator and discovered a series of periodic motions. Xu and Luo [57] also presented sequential period-($2m$-1) motions to chaos in the periodically forced van der Pol oscillator. In 2016, Guo and Luo [58, 59] employed such a method on a periodically forced pendulum. To know the inherent complex dynamics of the parametrically excited pendulum, Guo and Luo [60] investigated a parametrically excited pendulum system through the implicit mapping method. The parametric pendulum possesses different dynamical behaviors in the periodically forced pendulum. To help one understand dynamics of the parametric pendulum, the bifurcation trees of periodic motions in the parametrically excited pendulum will be presented herein.

Thus, in this short book, bifurcation dynamics of a parametric pendulum will be presented through the implicit mapping method. The implicit mapping method is a semi-analytical method for periodic motions in nonlinear dynamical systems, which will be presented in Chapter 2. Based on the semi-analytical method, periodic motions in the parametrically excited pendulum will be studied from the mapping structures, as presented in Chapter 3. In Chapter 4, bifurcation trees of periodic motions to chaos will be presented for period-1 static points, period-1, period-2, …, period-6, period-8, period-10, and period-12 motions to chaos. Harmonic frequency-amplitude characteristics for selected bifurcation trees will also be presented in Chapter 5. Both travelable and non-travelable periodic motions of the parametric pendulum will be illustrated from analytical predictions, as presented in Chapters 6 and 7. For non-travelable periodic motions, the corresponding harmonic amplitudes and phases will also be illustrated. For travelable periodic motions, the harmonic analysis of the travelable periodic motions will be based on the velocity rather than displacement because the initial and ending displacements do not equal each other. This book provides a clear skeleton of bifurcation trees of periodic motions to chaos in the parametric pendulums.

CHAPTER 2

A Semi-Analytical Method

Periodic motions in dynamical systems will be presented in this chapter. If a nonlinear system has a periodic motion with a period of $T = 2\pi/\Omega$, then such a periodic motion can be expressed by discrete points through discrete mappings of continuous dynamical systems. The method is stated through the following theorem. From Luo [48], we have the following theorem.

Theorem 2.1 *Consider a nonlinear dynamical system as*

$$\dot{\mathbf{x}} = \mathbf{f}(\mathbf{x}, t, \mathbf{p}) \in \mathcal{R}^n, \tag{2.1}$$

where $\mathbf{f}(\mathbf{x}, t, \mathbf{p})$ is a C^r-continuous nonlinear vector function ($r \geq 1$). If such a dynamical system has a periodic motion $\mathbf{x}(t)$ with finite norm $||\mathbf{x}||$ and period $T = 2\pi/\Omega$, there is a set of discrete time t_k ($k = 0, 1, \ldots, N$) with ($N \to \infty$) during one period T, and the corresponding solution $\mathbf{x}(t_k)$ and vector fields $\mathbf{f}(\mathbf{x}(t_k), t_k, \mathbf{p})$ are exact. Suppose a discrete node $\mathbf{x}_k$ is on the approximate solution of the periodic motion under $||\mathbf{x}(t_k) - \mathbf{x}_k|| \leq \varepsilon_k$ with a small $\varepsilon_k \geq 0$ and

$$||\mathbf{f}(\mathbf{x}(t_k), t_k, \mathbf{p}) - \mathbf{f}(\mathbf{x}_k, t_k, \mathbf{p})|| \leq \delta_k, \tag{2.2}$$

where a small $\delta_k \geq 0$. During a time interval $t \in [t_{k-1}, t_k]$, there is a general implicit mapping $P_k : \mathbf{x}_{k-1} \to \mathbf{x}_k$ ($k = 1, 2, \ldots, N$) as

$$\mathbf{x}_k = P_k \mathbf{x}_{k-1} \quad with \quad \mathbf{g}_k(\mathbf{x}_{k-1}, \mathbf{x}_k, \mathbf{p}) = \mathbf{0}, \quad k = 1, 2, \ldots, N, \tag{2.3}$$

where $\mathbf{g}_k$ is an implicit vector function. Consider a mapping structure as

$$P = P_N \circ P_{N-1} \circ \ldots \circ P_2 \circ P_1 : \mathbf{x}_0 \to \mathbf{x}_N; \tag{2.4}$$
$$with \quad P_k : \mathbf{x}_{k-1} \to \mathbf{x}_k \quad (k = 1, 2, \ldots, N).$$

For $\mathbf{x}_N = P\mathbf{x}_0$, if there is a set of nodes points $\mathbf{x}_k^*(k = 0, 1, \ldots, N)$ computed by

$$\mathbf{g}_k(\mathbf{x}_{k-1}^*, \mathbf{x}_k^*, \mathbf{p}) = 0, \quad (k = 1, 2, \ldots, N) \tag{2.5}$$
$$\mathbf{x}_0^* = \mathbf{x}_N^*,$$

then the points $\mathbf{x}_k^*$ ($k = 0, 1, \ldots, N$) are approximations of points $\mathbf{x}(t_k)$ of the periodic solution. In a neighborhood of $\mathbf{x}_k^*$, with $\mathbf{x}_k = \mathbf{x}_k^* + \Delta\mathbf{x}_k$, the linearized equation is given by

$$\Delta\mathbf{x}_k = DP_k \cdot \Delta\mathbf{x}_{k-1}$$
$$with \quad \mathbf{g}_k(\mathbf{x}_{k-1}^* + \Delta\mathbf{x}_{k-1}, \mathbf{x}_k^* + \Delta\mathbf{x}_k, \mathbf{p}) = 0 \tag{2.6}$$
$$(k = 1, 2, \ldots, N).$$

The resultant Jacobian matrix of the periodic motion is

$$DP_{k(k-1)\ldots1} = DP_k \cdot DP_{k-1} \cdot \ldots \cdot DP_1, \quad (k = 1, 2, \ldots, N);$$
$$DP \equiv DP_{N(N-1)\ldots1} = DP_N \cdot DP_{N-1} \cdot \ldots \cdot DP_1, \tag{2.7}$$

where

$$DP_k = \left[\frac{\partial \mathbf{x}_k}{\partial \mathbf{x}_{k-1}}\right]_{(\mathbf{x}_{k-1}^*, \mathbf{x}_k^*)} = -\left[\frac{\partial \mathbf{g}_k}{\partial \mathbf{x}_k}\right]^{-1}_{(\mathbf{x}_{k-1}^*, \mathbf{x}_k^*)} \left[\frac{\partial \mathbf{g}_k}{\partial \mathbf{x}_{k-1}}\right]_{(\mathbf{x}_{k-1}^*, \mathbf{x}_k^*)}. \tag{2.8}$$

The eigenvalues of DP and $DP_{k(k-1)\ldots1}$ for such a periodic motion are determined by

$$\left|DP_{k(k-1)\ldots1} - \bar{\lambda}\mathbf{I}_{n\times n}\right| = 0, \quad (k = 1, 2, \ldots, N);$$
$$|DP - \lambda\mathbf{I}_{n\times n}| = 0. \tag{2.9}$$

Thus, the stability and bifurcation of the periodic motion can be classified by the eigenvalues of $DP(\mathbf{x}_0^*)$ with

$$\left(\left[n_1^m, n_1^\circ\right] : \left[n_2^m, n_2^\circ\right] : [n_3, \kappa_3] : [n_4, \kappa_4] \,|n_5 : n_6 : [n_7, l, \kappa_7]\right), \tag{2.10}$$

where n_1 is the total number of real eigenvalues with magnitudes less than one ($n_1 = n_1^m + n_1^\circ$); n_2 is the total number of real eigenvalues with magnitude greater than one ($n_2 = n_2^m + n_2^\circ$); n_3 is the total number of real eigenvalues equal to $+1$; n_4 is the total number of real eigenvalues equal to -1; n_5 is the total pair number of complex eigenvalues with magnitudes less than one; n_6 is the total pair number of complex eigenvalues with magnitudes greater than one; and n_7 is the total pair number of complex eigenvalues with magnitudes equal to one.

(i) *If the magnitudes of all eigenvalues of DP are less than one, the approximate period-1 motion is stable.*

(ii) *If at least the magnitude of one eigenvalue of DP is greater than one, the approximate period-1 motion is unstable.*

(iii) *The boundaries between stable and unstable periodic motion give bifurcation and stability conditions.*

To explain how to approximate a periodic motion in an n-dimensional nonlinear dynamical system, consider an $n_1 \times n_2$ plane ($n_1 + n_2 = n$) as shown in Fig. 2.1. N-nodes of the periodic motion are chosen for an approximate solution with a certain accuracy $||\mathbf{x}(t_k) - \mathbf{x}_k|| \leq \varepsilon_k$ ($\varepsilon_k \geq 0$) and $|\mathbf{f}(\mathbf{x}(t_k), t_k, \mathbf{p}) - \mathbf{f}(\mathbf{x}_k, t_k, \mathbf{p})|| \leq \delta_k$ ($\delta_k \geq 0$). Letting $\delta = \max\{\delta_k\}_{k\in\{1,2,\ldots,N\}}$ and $\varepsilon = \max\{\varepsilon_k\}_{k\in\{1,2,\ldots,N\}}$ be small positive quantities prescribed, the periodic motion can be approximately described by a set of specific implicit mappings P_k with $\mathbf{g}_k(\mathbf{x}_{k-1}, \mathbf{x}_k, \mathbf{p}) = 0$ ($k = 1, 2, \ldots, N$) with a periodicity condition $\mathbf{x}_N = \mathbf{x}_0$. Based on the approximate mapping functions, the nodes of the trajectory of periodic motion are computed approximately, which is depicted by a solid curve. The exact solution of the periodic motion is described by a dashed

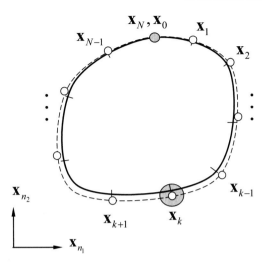

Figure 2.1: Period-1 motion with N-node points. Solid curve: numerical results with N-nodes marked by short lines, and dashed curve: expected exact results with N-nodes marked by symbols. The local shaded area is a small neighborhood of the exact solution at the kth node. The symbols are node points on the exact solution of the periodic motion.

curve. The node points on the periodic motion are depicted with short lines. The symbols are node points on the exact solution of the periodic motion. The discrete mapping P_k is developed from the differential equation. With the control of computational accuracy, the nodes of the periodic motion can be obtained with a good approximation.

From the stability and bifurcation analysis, the period-1 motion under period $T = 2\pi/\Omega$, based on the set of discrete implicit mapping P_k with $\mathbf{g}_k(\mathbf{x}_{k-1}, \mathbf{x}_k, \mathbf{p}) = 0$ $(k = 1, 2, \ldots, N)$, is stable or unstable. If the period-doubling bifurcation occurs, the periodic motion will become a periodic motion under period $T' = 2T$, and such a periodic motion is called a period-2 motion. Due to the period-doubling, $2N$ nodes of the period-2 motion will be employed to describe the period-2 motion. Thus, consider a mapping structure of the period-2 motion with $2N$ implicit mappings.

$$P = P_{2N} \circ P_{2N-1} \circ \ldots \circ P_2 \circ P_1 : \mathbf{x}_0 \to \mathbf{x}_{2N};$$
$$\text{with } P_k : \mathbf{x}_{k-1} \to \mathbf{x}_k \quad (k = 1, 2, \ldots, 2N).$$
$$(2.11)$$

For $\mathbf{x}_{2N} = P\mathbf{x}_0$, there is a set of points $\mathbf{x}_k^*$ $(k = 0, 1, \ldots, 2N)$ computed by the following implicit vector functions:

$$\mathbf{g}_k(\mathbf{x}_{k-1}^*, \mathbf{x}_k^*, \mathbf{p}) = 0, \quad (k = 1, 2, \ldots, 2N)$$
$$\mathbf{x}_0^* = \mathbf{x}_{2N}^*.$$
$$(2.12)$$

After period-doubling, the period-1 motion becomes period-2 motion. The node points increase to $2N$ points during two periods $(2T)$. The period-2 motion can be sketched in Fig. 2.2. The node points are determined through the discrete implicit mapping with a mathematical relation in Eq. (2.12). On the other hand,

$$T' = 2T = \frac{2(2\pi)}{\Omega} = \frac{2\pi}{\omega} \Rightarrow \omega = \frac{\Omega}{2}. \tag{2.13}$$

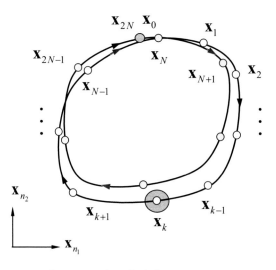

Figure 2.2: Period-2 motion with $2N$-nodes. Solid curve: numerical results. The symbols are node points on the periodic motion.

During the period of T', there is a periodic motion, which can be described by node points x_k $(k = 1, 2, \ldots, N')$. Since the period-1 motion is described by node points x_k $(k = 1, 2, \ldots, N)$ during the period T, due to $T' = 2T$, the period-2 motion can be described by $N' \geq 2N$ nodes. Thus, the corresponding mapping P_k is defined as

$$P_k : x_{k-1}^{(2)} \to x_k^{(2)} \quad (k = 1, 2, \ldots, 2N) \tag{2.14}$$

and

$$g_k(x_{k-1}^{(2)*}, x_k^{(2)*}, p) = 0, \quad (k = 1, 2, \ldots, 2N)$$
$$x_0^{(2)*} = x_{2N}^{(2)*}. \tag{2.15}$$

In general, for period $T' = mT$, there is a period-m motion which can be described by $N' \geq mN$. The corresponding mapping P_k is given by

$$P_k : x_{k-1}^{(m)} \to x_k^{(m)} \quad (k = 1, 2, \ldots, mN) \tag{2.16}$$

and

$$g_k^{(m)}(\mathbf{x}_{k-1}^{(m)*}, \mathbf{x}_k^{(m)*}, \mathbf{p}) = 0, \quad (k = 1, 2, \dots, mN)$$
$$\mathbf{x}_0^{(m)*} = \mathbf{x}_{mN}^{(m)*}. \tag{2.17}$$

From the above discussion, the period-m motion in a nonlinear dynamical system can be described through $(mN + 1)$ nodes for period mT. As in Luo [48], the corresponding theorem is presented as follows.

Theorem 2.2   *Consider a nonlinear dynamical system in Eq. (2.1). If such a dynamical system has a period-m motion $\mathbf{x}^{(m)}(t)$ with finite norm $||\mathbf{x}^{(m)}||$ and period mT ($T = 2\pi/\Omega$), there is a set of discrete time t_k ($k = 0, 1, \dots, mN$) with ($N \to \infty$) during m-periods (mT), and the corresponding solution $\mathbf{x}^{(m)}(t_k)$ and vector field $\mathbf{f}(\mathbf{x}^{(m)}(t_k), t_k, \mathbf{p})$ are exact. Suppose discrete node $\mathbf{x}_k^{(m)}$ is on the approximate solution of the periodic motion under $||\mathbf{x}^{(m)}(t_k) - \mathbf{x}_k^{(m)}|| \leq \varepsilon_k$ with a small $\varepsilon_k \geq 0$ and*

$$\left\| \mathbf{f}(\mathbf{x}^{(m)}(t_k), t_k, \mathbf{p}) - \mathbf{f}(\mathbf{x}_k^{(m)}, t_k, \mathbf{p}) \right\| \leq \delta_k \tag{2.18}$$

with a small $\delta_k \geq 0$. During a time interval $t \in [t_{k-1}, t_k]$, there is a general implicit mapping P_k : $\mathbf{x}_{k-1}^{(m)} \to \mathbf{x}_k^{(m)}$ ($k = 1, 2, \dots, mN$) as

$$\mathbf{x}_k^{(m)} = P_k \mathbf{x}_{k-1}^{(m)} \text{ with } \mathbf{g}_k(\mathbf{x}_{k-1}^{(m)}, \mathbf{x}_k^{(m)}, \mathbf{p}) = \mathbf{0}, \quad k = 1, 2, \dots, mN, \tag{2.19}$$

where $\mathbf{g}_k$ is an implicit vector function. Consider a mapping structure as

$$P = P_{mN} \circ P_{mN-1} \circ \dots \circ P_2 \circ P_1 : \mathbf{x}_0^{(m)} \to \mathbf{x}_{mN}^{(m)};$$
$$\text{with } P_k : \mathbf{x}_{k-1}^{(m)} \to \mathbf{x}_k^{(m)} \quad (k = 1, 2, \dots, mN). \tag{2.20}$$

For $\mathbf{x}_{mN}^{(m)} = P\mathbf{x}_0^{(m)}$, if there is a set of points $\mathbf{x}_k^{(m)*}$ ($k = 0, 1, \dots, mN$) computed by

$$\mathbf{g}_k(\mathbf{x}_{k-1}^{(m)*}, \mathbf{x}_k^{(m)*}, \mathbf{p}) = 0, \quad (k = 1, 2, \dots, mN)$$
$$\mathbf{x}_0^{(m)*} = \mathbf{x}_{mN}^{(m)*}, \tag{2.21}$$

then the points $\mathbf{x}_k^{(m)*}$ ($k = 0, 1, \dots, mN$) are approximations of points $\mathbf{x}^{(m)}(t_k)$ of the periodic solution. In a neighborhood of $\mathbf{x}_k^{(m)*}$, with $\mathbf{x}_k^{(m)} = \mathbf{x}_k^{(m)*} + \Delta\mathbf{x}_k^{(m)}$, the linearized equation is given by

$$\Delta\mathbf{x}_k^{(m)} = DP_k \cdot \Delta\mathbf{x}_{k-1}^{(m)}$$
$$\text{with } \mathbf{g}_k(\mathbf{x}_{k-1}^{(m)*} + \Delta\mathbf{x}_{k-1}^{(m)}, \mathbf{x}_k^{(m)*} + \Delta\mathbf{x}_k^{(m)}, \mathbf{p}) = 0 \tag{2.22}$$
$$(k = 1, 2, \dots, mN).$$

The resultant Jacobian matrices of the periodic motion are

$$
DP_{k(k-1)\dots 1} = DP_k \cdot DP_{k-1} \cdot \dots \cdot DP_1, \quad (k = 1, 2, \dots, mN);
$$
$$
DP \equiv DP_{mN(mN-1)\dots 1} = DP_{mN} \cdot DP_{mN-1} \cdot \dots \cdot DP_1, \tag{2.23}
$$

where

$$
DP_k = \left[\frac{\partial \mathbf{x}_k^{(m)}}{\partial \mathbf{x}_{k-1}^{(m)}} \right]_{\left(\mathbf{x}_{k-1}^{(m)*}, \mathbf{x}_k^{(m)*} \right)} = - \left[\frac{\partial \mathbf{g}_k}{\partial \mathbf{x}_k^{(m)}} \right]^{-1} \left[\frac{\partial \mathbf{g}_k}{\partial \mathbf{x}_{k-1}^{(m)}} \right]_{\left(\mathbf{x}_{k-1}^{(m)*}, \mathbf{x}_k^{(m)*} \right)}. \tag{2.24}
$$

The eigenvalues of $DP(\mathbf{x}_0^{(m)*})$ and $DP_{k(k-1)\dots 1}$ for such a periodic motion are determined by

$$
\left| DP_{k(k-1)\dots 1} - \bar{\lambda} \mathbf{I}_{n \times n} \right| = 0, \quad (k = 1, 2, \dots, mN);
$$
$$
|DP - \lambda \mathbf{I}_{n \times n}| = 0. \tag{2.25}
$$

Thus, the stability and bifurcation of the periodic motion can be classified by the eigenvalues of $DP(x_0^{(m)*})$ with

$$
\left(\left[n_1^m, n_1^\circ \right] : \left[n_2^m, n_2^\circ \right] : [n_3, \kappa_3] : [n_4, \kappa_4] \,|\, n_5 : n_6 : [n_7, l, \kappa_7] \right). \tag{2.26}
$$

(i)   *If the magnitudes of all eigenvalues of $DP^{(m)}$ are less than one (i.e., $|\lambda_i| < 1$, $i = 1, 2, \dots, n$), the approximate period-m solution is stable.*

(ii) *If at least the magnitude of one eigenvalue of $DP^{(m)}$ is greater than one (i.e., $|\lambda_i| > 1$, $i \in \{1, 2, \dots, n\}$), the approximate period-m solution is unstable.*

(iii) *The boundaries between stable and unstable period-m motions give bifurcation and stability conditions.*

CHAPTER 3

Discretization of a Parametric Pendulum

In this chapter, a semi-analytical method will be employed for periodic motions in the parametrically driven pendulum system through implicit discrete mappings. The implicit discrete mapping structures of periodic motions will be developed, and eigenvalue analysis will be used for the corresponding stability and bifurcation analysis.

3.1   IMPLICIT DISCRETE MAPPINGS

Consider a parametric pendulum shown in Fig. 3.1. The pendulum has a particle mass of m, connected with a massless bar of length l. Another end of the massless bar is moving harmonically with $Q \cos \Omega t$. The angular displacement of the pendulum is θ. The unit vectors on the horizontal and vertical directions are $\mathbf{i}$ and $\mathbf{j}$, respectively. Define a position vector

$$\mathbf{r} = l \sin \theta \mathbf{i} - (l \cos \theta + Q \cos \Omega t)\mathbf{j}, \qquad (3.1)$$

and the velocity vector is

$$\dot{\mathbf{r}} = -l \dot{\theta} \cos \theta \mathbf{i} - (l \dot{\theta} \sin \theta - Q \Omega \sin \Omega t)\mathbf{j}. \qquad (3.2)$$

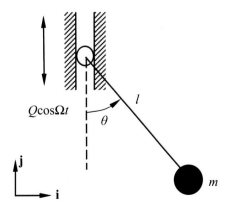

Figure 3.1: A mechanical model for a parametric pendulum.

The kinetic energy is

$$T = \frac{1}{2}m\dot{\mathbf{r}} \cdot \dot{\mathbf{r}}$$
$$= \frac{1}{2}m\left(l^2\dot{\theta}^2 - 2l\dot{\theta}Q\Omega \sin\theta \sin\Omega t + Q^2\Omega^2 \sin^2\Omega t\right). \tag{3.3}$$

Assume the damping force for the particle mass is related to the pure rotation only, i.e.,

$$\mathbf{F}_d = -cl\dot{\theta}\cos\theta\mathbf{i} - cl\dot{\theta}\sin\theta\mathbf{j}, \tag{3.4}$$

where c is damping coefficient, and the conservative force for the particle mass is

$$\mathbf{F}_c = -mg\mathbf{j}. \tag{3.5}$$

The resultant force for particle mass is

$$\mathbf{F} = \mathbf{F}_c + \mathbf{F}_d$$
$$= -cl\dot{\theta}\cos\theta\mathbf{i} - (mg + cl\dot{\theta}\sin\theta)\mathbf{j}. \tag{3.6}$$

Using the Lagrange equation, we have

$$\frac{\partial}{\partial t}\left(\frac{\partial T}{\partial \dot{\theta}}\right) - \frac{\partial T}{\partial \theta} = \mathbf{F} \cdot \frac{\partial \mathbf{r}}{\partial \theta}, \tag{3.7}$$

where

$$\frac{\partial T}{\partial \dot{\theta}} = ml^2\dot{\theta} - mlQ\Omega \sin\theta \sin\Omega t,$$
$$\frac{\partial T}{\partial \theta} = -ml\dot{\theta}Q\Omega \cos\theta \sin\Omega t,$$
$$\frac{\partial}{\partial t}\left(\frac{\partial T}{\partial \dot{\theta}}\right) = ml^2\ddot{\theta} - mlQ\Omega\dot{\theta} \cos\theta \sin\Omega t + mlQ\Omega^2 \sin\theta \cos\Omega t, \tag{3.8}$$
$$\frac{\partial \mathbf{r}}{\partial \theta} = l\cos\theta\mathbf{i} + l\sin\theta\mathbf{j},$$
$$\mathbf{F} \cdot \frac{\partial \mathbf{r}}{\partial \theta} = -cl^2\dot{\theta} - mgl \sin\theta.$$

From Eqs. (3.7) and (3.8), the equation of motion for the parametric pendulum is

$$ml^2\ddot{\theta} + cl^2\dot{\theta} + \left(mgl + mlQ\Omega^2 \cos\Omega t\right)\sin\theta = 0. \tag{3.9}$$

Letting

$$\delta = \frac{c}{m}, \quad \alpha = \frac{g}{l}, \quad Q_0 = \frac{Q\Omega^2}{l}, \quad x = \theta, \tag{3.10}$$

deformation of Eq. (3.9) gives a damped parametrically driven pendulum system as

$$\ddot{x} + \delta\dot{x} + (\alpha + Q_0 \cos \Omega t) \sin x = 0, \qquad (3.11)$$

where δ is damping coefficient, α is stiffness, and Q_0 and Ω are excitation amplitude and frequency, respectively. In phase space, such a system can be re-written as

$$\begin{aligned} \dot{x} &= y, \\ \dot{y} &= -\delta\dot{x} - (\alpha + Q_0 \cos \Omega t) \sin x. \end{aligned} \qquad (3.12)$$

Using a midpoint scheme for the time interval $t \in [t_{k-1}, t_k]$, the above system can be discretized to form an implicit map P_k $(k = 0, 1, 2, \ldots)$ as

$$P_k : (x_{k-1}, y_{k-1}) \to (x_k, y_k) \quad \Rightarrow \quad (x_k, y_k) = P_k (x_{k-1}, y_{k-1}), \qquad (3.13)$$

and the corresponding implicit relationships are

$$\begin{aligned} x_k &= x_{k-1} + \tfrac{1}{2}h (y_{k-1} + y_k), \\ y_k &= y_{k-1} - h\{\tfrac{1}{2}\delta (y_{k-1} + y_k) \\ &\quad + [\alpha + Q_0 \cos \Omega(t_{k-1} + \tfrac{1}{2}h)] \sin[\tfrac{1}{2}(x_{k-1} + x_k)]\}. \end{aligned} \qquad (3.14)$$

The above discretization experiences an accuracy of $O(h^3)$ for each step. To keep computational accuracy less than 10^{-8}, $h < 10^{-3}$ needs to be maintained.

3.2   PERIOD-1 MOTIONS

To analytically predict the periodic solutions in such a parametrically excited pendulum, consider a mapping structure as

$$P = \underbrace{P_N \circ P_{N-1} \circ \cdots \circ P_2 \circ P_1}_{N-actions} : (x_0, y_0) \to (x_N, y_N) \qquad (3.15)$$

with

$$\begin{aligned} P_1 &: (x_0, y_0) \to (x_1, y_1) \quad \Rightarrow \quad (x_1, y_1) = P_1 (x_0, y_0) \\ P_2 &: (x_1, y_1) \to (x_2, y_2) \quad \Rightarrow \quad (x_2, y_2) = P_2 (x_1, y_1) \\ &\;\;\vdots \\ P_{N-1} &: (x_{N-2}, y_{N-2}) \to (x_{N-1}, y_{N-1}) \quad \Rightarrow \quad (x_{N-1}, y_{N-1}) = P_{N-1} (x_{N-2}, y_{N-2}) \\ P_N &: (x_{N-1}, y_{N-1}) \to (x_N, y_N) \quad\;\; \Rightarrow \quad (x_N, y_N) = P_N (x_{N-1}, y_{N-1}). \end{aligned} \qquad (3.16)$$

For $t_k = t_0 + kh$ with given t_0 and h, from Eq. (3.14), the corresponding algebraic equations are

$$
\left.
\begin{aligned}
x_1 &= x_0 + \tfrac{1}{2}h(y_0 + y_1), \\
y_1 &= y_0 - h\{\tfrac{1}{2}\delta(y_0 + y_1) \\
&\quad + [\alpha + Q_0 \cos \Omega(t_0 + \tfrac{1}{2}h)]\sin[\tfrac{1}{2}(x_0 + x_1)]\}
\end{aligned}
\right\} \quad \text{for } P_1;
$$

$$\vdots$$

$$
\left.
\begin{aligned}
x_k &= x_{k-1} + \tfrac{1}{2}h(y_{k-1} + y_k), \\
y_k &= y_{k-1} - h\{\tfrac{1}{2}\delta(y_{k-1} + y_k) \\
&\quad + [\alpha + Q_0 \cos \Omega(t_{k-1} + \tfrac{1}{2}h)]\sin[\tfrac{1}{2}(x_{k-1} + x_k)]\}
\end{aligned}
\right\} \quad \text{for } P_k; \qquad (3.17)
$$

$$\vdots$$

$$
\left.
\begin{aligned}
x_N &= x_{N-1} + \tfrac{1}{2}h(y_{N-1} + y_N), \\
y_N &= y_{N-1} - h\{\tfrac{1}{2}\delta(y_{N-1} + y_N) \\
&\quad + [\alpha + Q_0 \cos \Omega(t_{N-1} + \tfrac{1}{2}h)]\sin[\tfrac{1}{2}(x_{N-1} + x_N)]\}
\end{aligned}
\right\} \quad \text{for } P_N.
$$

With periodicity conditions, we have

$$
(x_N, y_N) = (x_0 + 2l\pi, y_0), \quad l = 0, \pm 1, \pm 2, \ldots. \qquad (3.18)
$$

Definition 3.1   For a period-1 motion of dynamical system in Eq. (3.11) for N-nodes per period where $N = T/h$ with time step h, if

$$
x_k = x_{k+N} \quad \text{and} \quad y_k = y_{k+N}, \qquad (3.19)
$$

then such a period-1 motion is called the non-travelable period-m motion in the dynamical system.

Definition 3.2   For a period-1 motion of dynamical system in Eq. (3.11) for N-nodes per period where $N = T/h$ with time step h, if

$$
\begin{aligned}
&\text{mod}(x_k, 2\pi) = \text{mod}(x_{k+N}, 2\pi) \\
&\text{with } x_k \neq x_{k+N} \quad \text{and} \quad y_k = y_{k+N},
\end{aligned} \qquad (3.20)
$$

then such a period-1 motion is called the travelable period-1 motion in the dynamical system.

From Eqs. (3.17) and (3.18), values of nodes at the discretized parametric pendulum can be determined by $2(N + 1)$ equations. Such a periodic solution can be sketched in Fig. 3.2. The node points are depicted by the circular symbols, labeled by $\mathbf{x}_k = (x_k, y_k)^T$ ($k = 0, 1, 2, \ldots, N$), and the initial and final points are equal for periodicity. The mappings are depicted through the curves with arrows. Once the node points of the period-1 motion $\mathbf{x}_k^*$ $(k = 0, 1, 2, \ldots, N)$ are obtained, the stability of period-1 motion can be discussed by the corresponding Jacobian matrix. Consider a small perturbation in the neighborhood of $\mathbf{x}_k^*$, $\mathbf{x}_k = \mathbf{x}_k^* + \Delta \mathbf{x}_k$, $(k = 0, 1, 2, \ldots, N)$.

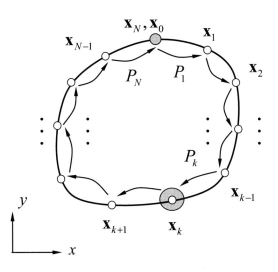

Figure 3.2: Period-1 motion with N-nodes of the parametric pendulum oscillator. The mapping structures are depicted through mappings with the arrowed curves. The circular symbols represent the node points of the period-1 motion.

For the mapping structure in Eq. (3.5), we have

$$\Delta \mathbf{x}_N = DP \Delta \mathbf{x}_0 = \underbrace{DP_N \cdot DP_{N-1} \cdot \ldots \cdot DP_2 \cdot DP_1}_{N\text{-multiplication}} \Delta \mathbf{x}_0 \qquad (3.21)$$

with

$$\Delta \mathbf{x}_1 = DP_1 \Delta \mathbf{x}_0 \equiv \left[\frac{\partial \mathbf{x}_1}{\partial \mathbf{x}_0} \right]_{(\mathbf{x}_1^*, \mathbf{x}_0^*)} \Delta \mathbf{x}_0$$

$$\Delta \mathbf{x}_2 = DP_2 \Delta \mathbf{x}_1 \equiv \left[\frac{\partial \mathbf{x}_2}{\partial \mathbf{x}_1} \right]_{(\mathbf{x}_2^*, \mathbf{x}_1^*)} \Delta \mathbf{x}_1,$$

$$\vdots \qquad\qquad\qquad\qquad\qquad\qquad (3.22)$$

$$\Delta \mathbf{x}_{N-1} = DP_{N-1} \Delta \mathbf{x}_{N-2} \equiv \left[\frac{\partial \mathbf{x}_{N-1}}{\partial \mathbf{x}_{N-2}} \right]_{(\mathbf{x}_{N-1}^*, \mathbf{x}_{N-2}^*)} \Delta \mathbf{x}_{N-2},$$

$$\Delta \mathbf{x}_N = DP_N \Delta \mathbf{x}_{N-1} \equiv \left[\frac{\partial \mathbf{x}_N}{\partial \mathbf{x}_{N-1}} \right]_{(\mathbf{x}_N^*, \mathbf{x}_{N-1}^*)} \Delta \mathbf{x}_{N-1};$$

where

$$DP_k = \left[\frac{\partial \mathbf{x}_k}{\partial \mathbf{x}_{k-1}}\right]_{(\mathbf{x}_k^*, \mathbf{x}_{k-1}^*)} = \left[\begin{array}{cc} \dfrac{\partial x_k}{\partial x_{k-1}} & \dfrac{\partial x_k}{\partial y_{k-1}} \\ \\ \dfrac{\partial y_k}{\partial x_{k-1}} & \dfrac{\partial y_k}{\partial y_{k-1}} \end{array}\right]_{(\mathbf{x}_k^*, \mathbf{x}_{k-1}^*)} \quad \text{for } k = 1, 2, \ldots, N \qquad (3.23)$$

and

$$\frac{\partial x_k}{\partial x_{k-1}} = 1 + \frac{1}{2}h\frac{\partial y_k}{\partial x_{k-1}},$$

$$\frac{\partial x_k}{\partial y_{k-1}} = \frac{1}{2}h(1 + \frac{\partial y_k}{\partial y_{k-1}}),$$

$$\frac{\partial y_k}{\partial x_{k-1}} = -\frac{1}{2}h\left\{(1 + \frac{\partial x_k}{\partial x_{k-1}})[\alpha + Q_0 \cos \Omega(t_{k-1} + \frac{1}{2}h)]\right.$$

$$\left. \times \cos[\frac{1}{2}(x_{k-1} + x_k)] + \delta\frac{\partial y_k}{\partial x_{k-1}}\right\}, \qquad (3.24)$$

$$\frac{\partial y_k}{\partial y_{k-1}} = 1 - \frac{1}{2}h\left\{[\alpha + Q_0 \cos \Omega(t_{k-1} + \frac{1}{2}h)]\right.$$

$$\left. \times \cos[\frac{1}{2}(x_{k-1} + x_k)]\frac{\partial x_k}{\partial y_{k-1}} + \delta(1 + \frac{\partial y_k}{\partial y_{k-1}})\right\}.$$

To measure stability and bifurcation of the period-1 motion, the eigenvalues are computed by

$$|DP - \lambda I| = 0, \qquad (3.25)$$

where

$$DP = \left[\frac{\partial \mathbf{x}_N}{\partial \mathbf{x}_0}\right]_{(\mathbf{x}_N^*, \mathbf{x}_{N-1}^*, \ldots, \mathbf{x}_0^*)} = DP_N \cdot \ldots \cdot DP_2 \cdot DP_1 = \prod_{k=N}^{1}\left[\frac{\partial \mathbf{x}_k}{\partial \mathbf{x}_{k-1}}\right]_{(\mathbf{x}_k^*, \mathbf{x}_{k-1}^*)}. \qquad (3.26)$$

For stability analysis, (i) if $|\lambda_i| < 1$ for $(i = 1, 2)$, the periodic motion is stable, and (ii) if $|\lambda_i| > 1$ for $(i \in \{1, 2\})$, the periodic motion is unstable.

For bifurcation analysis, if $\lambda_i = -1$ and $|\lambda_j| < 1$ for $(i, j \in \{1, 2\}$ and $j \neq i)$, the period-doubling bifurcation (PD) of periodic motion occurs. If $\lambda_i = 1$ and $|\lambda_j| < 1$ for $(i, j \in \{1, 2\}$ and $j \neq i)$, the saddle-node bifurcation (SN) of the periodic motion occurs. If $\lambda_{1,2}$ is a pair of complex eigenvalues and $|\lambda_{1,2}| = 1$, the Neimark bifurcation (NB) of the periodic motion occurs.

3.3   PERIOD-m MOTIONS

From Luo [48], for a period-m periodic motion in the damped parametrically excited pendulum, a set of discrete time t_k $(k = 0, 1, \ldots, mN)$ with $N \gg 1$ can be determined in m-periods.

Based on the discrete time, a set of discrete points $\mathbf{x}_k^{(m)} = (x_k^{(m)}, y_k^{(m)})$ can be determined on the trajectory of such a period-m motion. Implicit discrete maps are developed as

$$P_k : \left(x_{k-1}^{(m)}, y_{k-1}^{(m)}\right) \to \left(x_k^{(m)}, y_k^{(m)}\right) \quad \Rightarrow \quad \left(x_k^{(m)}, y_k^{(m)}\right) = P_k\left(x_{k-1}^{(m)}, y_{k-1}^{(m)}\right)$$

$$(k = 1, 2, \ldots, mN).$$

(3.27)

The period-m motion can then be represented by a discrete mapping structure, i.e.,

$$P = \underbrace{P_{mN} \circ P_{mN-1} \circ \ldots \circ P_2 \circ P_1}_{mN-\text{actions}} : \left(x_0^{(m)}, y_0^{(m)}\right) \to \left(x_{mN}^{(m)}, y_{mN}^{(m)}\right).$$

(3.28)

From the implicit relationship in Eq. (3.14), a set of governing equations can be obtained for each mapping as

$$\left.\begin{aligned}
x_k^{(m)} &= x_{k-1}^{(m)} + \tfrac{1}{2}h(y_{k-1}^{(m)} + y_k^{(m)}), \\
y_k^{(m)} &= y_{k-1}^{(m)} - h\left\{\tfrac{1}{2}\delta(y_{k-1}^{(m)} + y_k^{(m)}) \right. \\
&\quad \left. +(\alpha + Q_0 \cos \Omega(t_{k-1} + \tfrac{1}{2}h)) \sin\left[\tfrac{1}{2}(x_{k-1}^{(m)} + x_k^{(m)})\right]\right\}
\end{aligned}\right\} \text{ for } P_k$$

$$(k = 1, 2, \ldots, mN).$$

(3.29)

The corresponding periodicity is guaranteed by the following conditions:

$$(x_{mN}^{(m)}, y_{mN}^{(m)}) = (x_0^{(m)} + 2l\pi, y_0^{(m)}), \quad l = 0, \pm 1, \pm 2, \ldots; \; m = 1, 2, \ldots.$$

(3.30)

According to Luo and Guo [36], two types of periodic motions can be observed in such a pendulum system, the non-travelable and travelable periodic motions are defined as follows.

Definition 3.3   For a period-m motion of dynamical system in Eq. (3.11) for N-nodes per period where $N = T/h$ with time step h, if

$$x_k = x_{k+mN} \quad \text{and} \quad y_k = y_{k+mN},$$

(3.31)

then such a period-m motion is called the non-travelable period-m motion in the dynamical system.

Definition 3.4   For a period-m motion of dynamical system in Eq. (3.11) for N-nodes per period where $N = T/h$ with time step h, if

$$\text{mod}\,(x_k, 2\pi) = \text{mod}\,(x_{k+mN}, 2\pi)$$

$$\text{with} \; x_k \neq x_{k+mN} \quad \text{and} \quad y_k = y_{k+mN},$$

(3.32)

then such a period-m motion is called the travelable period-m motion in the dynamical system.

Using Eqs. (3.29) and (3.30), all the node points can be determined by solving the $2(mN + 1)$ equations as $\mathbf{x}_k^{(m)*}$. The corresponding stability and bifurcation of such period-m motion can be discussed through the Jacobian matrix. For a small perturbation in vicinity of $\mathbf{x}_k^{(m)*}$, $\mathbf{x}_k^{(m)} = \mathbf{x}_k^{(m)*} + \Delta\mathbf{x}_k^{(m)}$, $(k = 0, 1, 2, \ldots, mN)$,

$$\Delta\mathbf{x}_{mN}^{(m)} = DP\Delta\mathbf{x}_0^{(m)} = \underbrace{DP_{mN} \cdot DP_{mN-1} \cdot \ldots \cdot DP_2 \cdot DP_1}_{mN\text{-multiplication}} \Delta\mathbf{x}_0^{(m)}. \tag{3.33}$$

For each individual map in the mapping structure there is

$$\Delta\mathbf{x}_k^{(m)} = DP_k \Delta\mathbf{x}_{k-1}^{(m)} \equiv \left[\frac{\partial\mathbf{x}_k^{(m)}}{\partial\mathbf{x}_{k-1}^{(m)}}\right]_{\left(\mathbf{x}_k^{(m)*}, \mathbf{x}_{k-1}^{(m)*}\right)} \Delta\mathbf{x}_{k-1}^{(m)} \tag{3.34}$$

$$(k = 1, 2, \ldots, mN),$$

where the Jacobian matrix for each individual map is

$$DP_k = \left[\frac{\partial\mathbf{x}_k^{(m)}}{\partial\mathbf{x}_{k-1}^{(m)}}\right]_{\left(\mathbf{x}_k^{(m)*}, \mathbf{x}_{k-1}^{(m)*}\right)} = \begin{bmatrix} \dfrac{\partial x_k^{(m)}}{\partial x_{k-1}^{(m)}} & \dfrac{\partial x_k^{(m)}}{\partial y_{k-1}^{(m)}} \\[3mm] \dfrac{\partial y_k^{(m)}}{\partial x_{k-1}^{(m)}} & \dfrac{\partial y_k^{(m)}}{\partial y_{k-1}^{(m)}} \end{bmatrix}_{\left(\mathbf{x}_k^{(m)*}, \mathbf{x}_{k-1}^{(m)*}\right)} \tag{3.35}$$

$$\text{for} \quad k = 1, 2, \ldots, mN.$$

Thus, the resultant Jacobian matrix for such a period-m motion is

$$DP = \prod_{k=mN}^{1} \left[\frac{\partial\mathbf{x}_k^{(m)}}{\partial\mathbf{x}_k^{(m)}}\right]_{\left(\mathbf{x}_k^{(m)*}, \mathbf{x}_{k-1}^{(m)*}\right)}. \tag{3.36}$$

The corresponding eigenvalues for the period-m motion are computed by

$$|DP - \lambda\mathbf{I}| = 0. \tag{3.37}$$

The stability and bifurcation conditions for period-m motions are similar to the period-1 motions. If a period-m motion is determined by $x_k = x_{k+mN}$ with $y_k = y_{k+mN}$, such a period-m motion is called a non-travelable period-m motion. On the other hand, if a period-m motion is determined by $\text{mod}(x_k, 2\pi) = \text{mod}(x_{k+mN}, 2\pi)$ with $y_k = y_{k+mN}$ but $x_k \neq x_{k+mN}$, such a period-m motion is called a travelable period-m motion.

CHAPTER 4

Bifurcation Trees

In this chapter, the possible bifurcation trees are presented for analytical predictions of the routes of different periodic motions to chaos in the parametrically excited pendulum. The stability and bifurcations of periodic motions are also illustrated through eigenvalue analysis. The solid and dashed curves represent the stable and unstable motions, respectively. The black and red colors are for paired asymmetric motions. The acronyms "SN" and "PD" represent the saddle-node and period-doubling bifurcations, respectively. The symmetric and asymmetric periodic motions are labeled by "$\underline{S}$" and "$\underline{A}$", respectively. All bifurcations trees are predicted with varying excitation frequency Ω. Other parameters are chosen as

$$\alpha = 4.0, \qquad \delta = 0.1, \qquad Q_0 = 5.0. \qquad (4.1)$$

4.1 PERIOD-1 STATIC POINTS TO CHAOS

There are two different types of bifurcation trees for period-1 motions to chaos: period-1 static points to chaos and asymmetric period-1 motions to chaos. In order to demonstrate the period-1 static points to chaos, the analytical bifurcation trees of period-1 static points to period-2 motions varying with excitation frequency Ω is illustrated in Fig. 4.1. The bifurcation trees are illustrated for $\Omega \in (0.7, 6.7)$, where solid and dashed curves depict stable and unstable motions, respectively. In Figs. 4.1a,b, presented are the predicted displacement and velocity of periodic nodes $\mathrm{mod}(x_{\mathrm{mod}(k,N)}, 2\pi)$ and $y_{\mathrm{mod}(k,N)}$ for $\mathrm{mod}(k, N) = 0$, respectively. The period-1 and period-2 motions are labeled through P-1 and P-2, respectively. In the bifurcation trees, the asymmetric period-1 motions are generated by saddle-node bifurcations of period-1 static point. The asymmetric period-2 motions appear through period-doubling bifurcations of the asymmetric period-1 motions, and the symmetric period-2 motions appear from period-doubling bifurcations of the period-1 static point. The cascaded period-doubling bifurcations continue, and period-4, period-8, period-16,… motions and eventually to chaos can be introduced. The real parts and magnitudes of eigenvalues for all periodic motions are illustrated in Figs. 4.1c,d, respectively. For a better illustration of bifurcations trees, two zoomed views of displacement and velocity for $\Omega \in (0.7, 1.6)$ are shown in Figs. 4.1e,f. The global bifurcation tree becomes more and more complicated for excitation frequency from $\Omega = 6.7$ to 0.7. The detailed views for node displacements and velocities are presented in Figs. 4.2g–j. The bifurcation structures can be clearly observed. The period-1 static points exist throughout the entire range. Such static-point motion has period-doubling and saddle-node bifurcations alternatively and keeps switch-

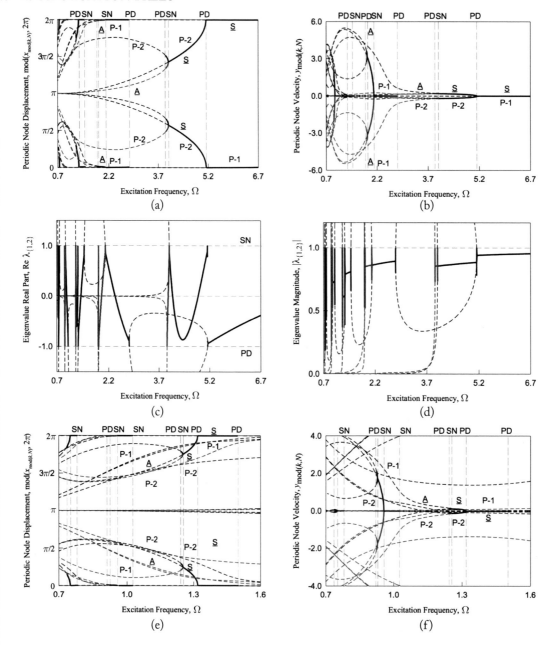

Figure 4.1: Bifurcation tree of period-1 static point to chaos with $\Omega \in (0.7, 6.7)$. Global view: (a) node displacement $x_{\mathrm{mod}(k,N)}$, (b) node velocity $y_{\mathrm{mod}(k,N)}$, (c) real part of eigenvalues, (d) eigenvalue magnitudes. Zoomed view with $\Omega \in (0.7, 1.6)$: (e) node displacement $\mathrm{mod}(x_{\mathrm{mod}(k,N)}, 2\pi)$, (f) node velocity $y_{\mathrm{mod}(k,N)}$.

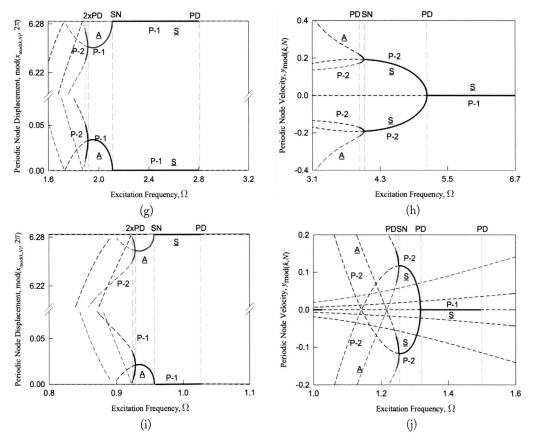

Figure 4.2: Detailed structures: (g) node displacement $x_{\mathrm{mod}(k,N)}$, (h) node velocity $y_{\mathrm{mod}(k,N)}$, (i) node displacement $x_{\mathrm{mod}(k,N)}$, (j) node velocity $y_{\mathrm{mod}(k,N)}$. ($\alpha = 4.0$, $\delta = 0.1$, $Q_0 = 5.0$.)

ing its stability between stable and unstable. For $\Omega \in (5.1380,\ 6.7000)$, the period-1 static point is stable. A period-doubling bifurcation of such period-1 static point occurs at $\Omega \approx 5.1380$, and symmetric period-2 motion is generated while the period-1 static point is from stable to unstable. The symmetric period-2 motion is stable for $\Omega \in (4.0130, 5.1380)$. Such a motion has a saddle-node bifurcation at $\Omega \approx 4.0130$ where the symmetric period-2 motion becomes unstable and asymmetric period-2 motions are introduced. Such asymmetric period-2 motions are stable for $\Omega \in (3.9391, 4.0130)$. Further cascaded period-doubling bifurcation occurs at $\Omega \approx 3.9391$ for asymmetric period-4, period-8... motions to chaos. Such period-4, period-8... motions are not illustrated herein because of the very tiny ranges of stable motions.

As excitation frequency decreases, the period-1 static point is from unstable to stable at the period-doubling bifurcation at $\Omega \approx 2.8039$. The motions induced by such period-doubling bifurcation are not presented owing to the tiny ranges of stable motions. Such period-1 static point stays stable for the range $\Omega \in (2.1080, 2.8039)$. At $\Omega \approx 2.1080$, a saddle-node bifurcation occurs where the period-1 static point becomes unstable again and a pair of asymmetric period-1 motions are generated. The induced asymmetric period-1 motions exist for $\Omega \in (1.9129, 2.1080)$, which are not static anymore. Such asymmetric period-1 motions maintain stable until the period-doubling bifurcation at $\Omega \approx 1.9129$. Thus, the asymmetric period-1 motions become unstable and asymmetric period-2 motions are induced. Such period-2 motions are stable for the range of $\Omega \in (1.8949, 1.9129)$. Further cascaded period-doubling bifurcations occur at $\Omega \approx 1.8949$. The induced period-4, period-8... motions are not presented owing to the tiny ranges of the sable motions.

With excitation frequency decrease, at $\Omega \approx 1.4989$, a saddle-node bifurcation of the period-1 static-point occurs and the period-1 static-point is from unstable to stable again. The range of the stable period-1 static-point is for $\Omega \in (1.3182, 1.4989)$. The period-doubling bifurcation of $\Omega \approx 1.3182$ switches the period-1 static point back to unstable while a symmetric period-2 motion is produced. Such symmetric period-2 motion is stable for $\Omega \in (1.2523, 1.3182)$. Such period-2 motion becomes unstable through the saddle-node bifurcation of $\Omega \approx 1.2523$, and a pair of asymmetric period-2 motions are introduced with a stable range of $\Omega \in (1.2452, 1.2523)$. At $\Omega \approx 1.2452$, a cascaded period-doubling bifurcation occurs for period-4, period-8... motions to chaos, as presented in Figs. 4.1e,f. As the excitation frequency further decreases, two more stable ranges for the period-1 static-point can be observed. The first one exists for $\Omega \in (0.9567, 1.0262)$, which connects to the unstable period-1 static-points through the period-doubling bifurcation at $\Omega \approx 1.0262$ and the saddle-node bifurcation at $\Omega \approx 0.9567$. Stable asymmetric period-1 motions are introduced for $\Omega \in (0.9281, 0.9567)$ through the saddle-node bifurcation at $\Omega \approx 0.9567$. Such period-1 motions become unstable with a birth of period-2 motions after the period-doubling bifurcation at $\Omega \approx 0.9281$. The period-2 motions are stable for $\Omega \in (0.9247, 0.9281)$ until the next period-doubling bifurcations at $\Omega \approx 0.9247$. The last stable period-1 static-point exists for $\Omega \in (0.7498, 0.7819)$, which connects to the unstable period-1 static-point through the saddle-node bifurcation at $\Omega \approx 0.7819$ and the period-doubling bifurcation at $\Omega \approx 0.7498$. Symmetric period-2 motion is introduced at $\Omega \approx 0.7498$ through a period-doubling bifurcation, and further cascaded period-doubling bifurcations to chaos are observed at $\Omega \approx 0.7357$.

4.2   PERIOD-1 AND PERIOD-3 MOTIONS TO CHAOS

The non-static period-1 and period-3 motions to chaos are demonstrated through the bifurcation trees of period-1 to period-4 motions, as shown in Fig. 4.3 for $\Omega \in (1.0, 5.2)$. Such period-1 motions are a pair of non-static asymmetric periodic motions not from the saddle-node bifurcations of symmetric motions. There are two branches of travelable asymmetric period-1 motions, as shown in Figs. 4.3a,b. The first branch of the asymmetric period-1 motion is stable for the range of $\Omega \in (3.9640, 5.2000)$ and becomes unstable at the period-doubling bifurcation of

$\Omega \approx 3.9640$. Such a period-doubling bifurcation also leads to a pair of asymmetric period-2 motions which are stable for $\Omega \in (3.6691, 3.9640)$. The period-doubling bifurcations at $\Omega \approx 3.6691$ cause a transition of stable to unstable period-2 motions and a pair of stable asymmetric period-4 motions for $\Omega \in (3.6405, 3.6691)$. Further, a cascaded period-doubling bifurcation occurs at $\Omega \approx 3.6405$, for period-8, period-16… motions to chaos. The second branch of the non-static period-1 motions exist for $\Omega \in (1.1587, 1.2441)$. Such period-1 motions are asymmetric and are independent of a symmetric period-1 motion. At $\Omega \approx 1.2441$, the saddle-node bifurcation is the onset of the asymmetric period-1 motions. At $\Omega \approx 1.1587$, the period-doubling bifurcation of the period-1 motion is for onset of asymmetric period-2 motions. The stable period-2 motion is for $\Omega \in (1.1378, 1.1587)$ until next cascaded period-doubling bifurcation of $\Omega \approx 1.1378$.

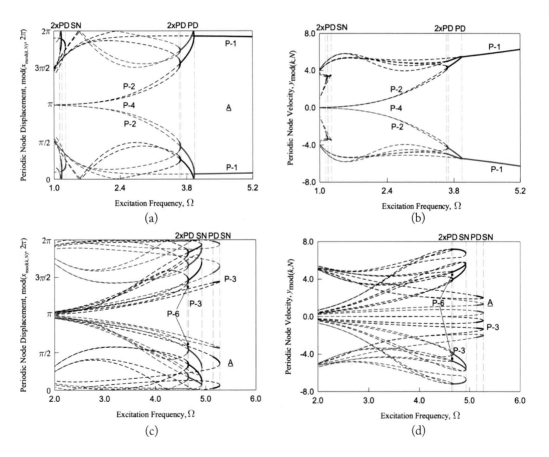

Figure 4.3: Period-1 motions to chaos with $\Omega \in (1.0, 5.2)$: (a) displacement $x_{\text{mod}(k,N)}$, (b) velocity $y_{\text{mod}(k,N)}$. Period-3 motions to chaos with $\Omega \in (2.0, 6.0)$: (c) displacement $\text{mod}(x_{\text{mod}(k,N)}, 2\pi)$, (d) velocity $y_{\text{mod}(k,N)}$. ($\alpha = 4.0$, $\delta = 0.1$, $Q_0 = 5.0$.)

For the parametric pendulum system with the same parameters, bifurcation trees of period-3 to period-6 motions are presented in Figs. 4.3c,d. There are also two branches of independent period-3 bifurcation trees to chaos, which are based on asymmetric period-3 motions. On the first branch, asymmetric period-3 motions are non-travelable and stable for $\Omega \in (5.1450, 5.2782)$. At $\Omega \approx 5.2782$, a saddle-node bifurcation is for the onset of the asymmetric period-3 motion. At $\Omega \approx 5.1450$, the period-doubling bifurcation of the asymmetric period-3 motions generates the birth of the period-6 motion. Continually, the cascaded period-doubling bifurcations will generate period-6, period-12... motions and eventually to chaos. The second branch of bifurcation tree of the period-3 motion to chaos has a pair of stable asymmetric period-3 motions in $\Omega \in (4.6575, 4.9248)$, which is travelable. The saddle-node bifurcation occur at $\Omega \approx 4.9248$ for the onset of the asymmetric period-3 motions. At $\Omega \approx 4.6675$, the period-doubling bifurcation of the period-3 motions produces stable asymmetric period-6 motions for $\Omega \in (4.6310, 4.6575)$. A next period-doubling bifurcation of $\Omega \approx 4.6310$ is for period-12, period-24... motions to chaos.

4.3   INDEPENDENT PERIOD-2 MOTIONS TO CHAOS

The bifurcation trees of independent period-2 motions to chaos are presented in Fig. 4.4. These bifurcation trees of period-2 motions are non-travelable. Such bifurcation trees are not from period-1 motions. The global views of such bifurcation trees are illustrated in Figs. 4.4a,b for $\Omega \in (0.5, 2.1)$. A zoomed view for the period-2 motions to chaos is provided in Figs. 4.4c,d for $\Omega \in (0.5, 1.0)$. Three branches of independent bifurcation trees of period-2 motion to chaos are observed. The period-2 motions are from symmetric to asymmetric. Starting from $\Omega = 2.1$, the first branch has a stable symmetric period-2 motion for $\Omega \in (1.7705, 2.0380)$. At $\Omega \approx 2.0380$, the saddle-node bifurcation is for onset of symmetric period-2 motion. At $\Omega \approx 1.7705$, a saddle-node bifurcation is for the onset of asymmetric period-2 motion. The asymmetric period-2 motions are stable for $\Omega \in (1.7217, 1.7705)$. A period-doubling bifurcation of the asymmetric period-2 motion at $\Omega \approx 1.7217$ is for period-4, period-8... motions to chaos. With excitation frequency decrease, the second bifurcation tree of the period-2 motion to chaos exists. The symmetric period-2 motion is stable for $\Omega \in (0.8656, 0.9084)$. Such symmetric period-2 motion has two saddle-node bifurcations at both ends. At $\Omega \approx 0.9084$, the saddle-node bifurcation is for onset of the symmetric period-2 motion. At $\Omega \approx 0.8656$, the saddle-node bifurcation is for a pair of asymmetric period-2 motions. Such period-2 motions are stable for $\Omega \in (0.8526, 0.8656)$ until the next period-doubling bifurcations at $\Omega \approx 0.8526$. Similarly, the third bifurcation tree of period-2 motion to chaos has a stable symmetric period-2 motion in $\Omega \in (0.5768, 0.5975)$. The saddle-node bifurcation for onset of the symmetric period-2 motion is at $\Omega \approx 0.5975$, while the saddle-node bifurcation for the asymmetric period-2 motions is at $\Omega \approx 0.5768$. The stable range for the asymmetric period-2 motions is $\Omega \in (0.5687, 0.5768)$ with the next period-doubling bifurcation at $\Omega \approx 0.5687$.

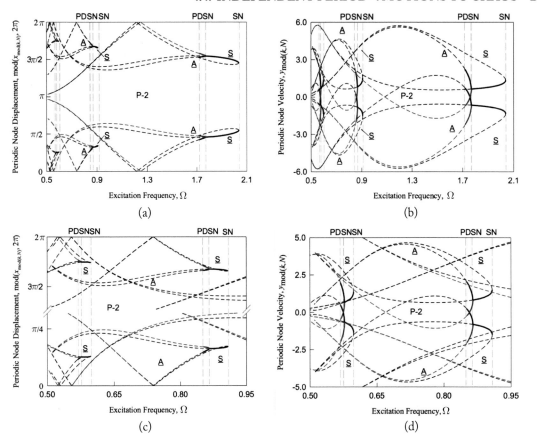

Figure 4.4: Bifurcation tree of period-2 motions to chaos varying with excitation frequency Ω. Global view $\Omega \in (0.5, 2.1)$: (a) node displacement $x_{\mathrm{mod}(k,N)}$, (b) node velocity $y_{\mathrm{mod}(k,N)}$. A zoomed view $\Omega \in (0.5, 1.0)$: (c) periodic node displacement $\mathrm{mod}(x_{\mathrm{mod}(k,N)}, 2\pi)$, (d) periodic node velocity $y_{\mathrm{mod}(k,N)}$. ($\alpha = 4.0$, $\delta = 0.1$, $Q_0 = 5.0$.)

4.4   INDEPENDENT PERIOD-4 MOTIONS TO CHAOS

Two types of different independent bifurcation trees are for symmetric or asymmetric period-4 motions to chaos. The two types of bifurcation trees of period-4 motions are independent of each other. The symmetric and asymmetric period-4 motions are also not from period-2 motions through period-doubling bifurcations.

Three branches of independent symmetric period-4 motions are presented in Figs. 4.5a,b for $\Omega \in (2.0, 8.0)$. On the first branch, the symmetric period-4 motion is stable for $\Omega \in (6.2556, 7.5379)$ with a saddle-node bifurcation at $\Omega \approx 7.5379$ for the onset. The saddle-node bifurcation at $\Omega \approx 6.2556$ is for onset of a pair of asymmetric period-4 motions. Such asymmet-

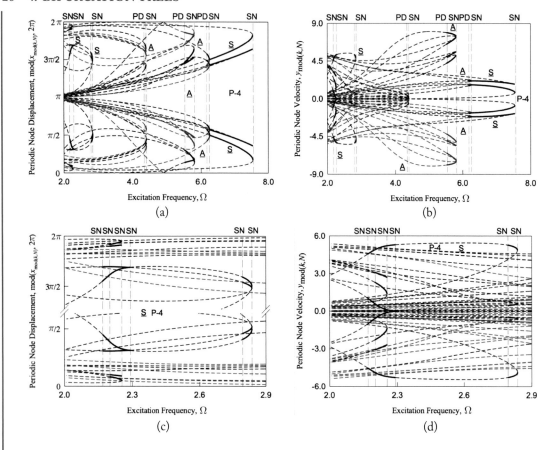

Figure 4.5: Bifurcation trees of symmetric period-4 motions to chaos varying with excitation frequency Ω. A global view $\Omega \in (2.0, 8.0)$: (a) node displacement $x_{\mathrm{mod}(k,N)}$, (b) node velocity $y_{\mathrm{mod}(k,N)}$. A zoomed view of $\Omega \in (2.0, 3.0)$: (c) periodic node displacement $\mathrm{mod}(x_{\mathrm{mod}(k,N)}, 2\pi)$, (d) periodic node velocity $y_{\mathrm{mod}(k,N)}$. ($\alpha = 4.0$, $\delta = 0.1$, $Q_0 = 5.0$.)

ric period-4 motions are stable for $\Omega \in (6.1788, 6.2556)$ until the cascaded period-doubling bifurcation at $\Omega \approx 6.1788$, where the asymmetric period-4 motions become unstable, and period-8, period-16… motions can be generated. The second independent symmetric period-4 motions to chaos has two stable ranges of $\Omega \in (2.1696, 2.2919)$ and $\Omega \in (2.7951, 2.8371)$ for the symmetric period-4 motions. At $\Omega \approx 2.8371$, a saddle-node bifurcation is for the onset of period-4 motion. At $\Omega \approx 2.2919$ and $\Omega \approx 2.7951$, a pair of saddle-node bifurcations are for unstable symmetric period-4 motion connecting with the two stable motions together. The two saddle-node bifurcations also produce asymmetric period-4 motions. However, such asymmetric period-4 motions are not illustrated herein due to the very tiny stable ranges. Finally, at

$\Omega \approx 2.1696$, another saddle-node bifurcation occurs, where the symmetric period-4 motions switches from stable to unstable, and asymmetric period-4 motions are introduced. Again, such asymmetric period-4 motions are not illustrated herein. On the third bifurcation tree, the independent symmetric period-4 motion is stable for $\Omega \in (2.2013, 2.2543)$, which overlay with part of the second bifurcation tree. At $\Omega \approx 2.2543$, a saddle-node bifurcation is for the onset of symmetric period-4 motion with a stable to unstable switching. At $\Omega \approx 2.2013$, the saddle-node bifurcation introduces asymmetric period-4 motions while the symmetric period-4 motions become unstable. A more detailed presentation is provided through the zoomed view for $\Omega \in (2.0, 3.0)$ in Figs. 4.5c,d. The aforementioned three bifurcation trees of period-4 motions to chaos are non-travelable. However, the fourth branch of asymmetric period-4 motions to chaos is travelable for $\Omega \in (0.0, 5.8029)$. The onset of such asymmetric period-4 motions occurs at $\Omega \approx 5.8029$, and the stable asymmetric period-4 motions exist for $\Omega \in (5.5327, 5.8029)$. At $\Omega \approx 5.5237$, the period-doubling bifurcation of the asymmetric period-4 motion occurs for the asymmetric period-8 motion, which will not be presented herein.

4.5   PERIOD-5 AND PERIOD-6 MOTIONS TO CHAOS

The bifurcation trees for travelable asymmetric period-5 motions to chaos are presented in Figs. 4.6a,b for $\Omega \in (2.0, 6.5)$. There are two bifurcation trees of period-5 motions to chaos. The paired period-5 motions are asymmetric. On the first bifurcation tree, the asymmetric period-5 motion is stable for $\Omega \in (6.0804, 6.1537)$. The onset of such asymmetric period-5 motions is at the saddle-node bifurcation of $\Omega \approx 6.1537$. At $\Omega \approx 6.0804$, period-doubling bifurcation of the asymmetric period-5 motion introduces period-10, period-20... motions which are not illustrated due to the tiny stable ranges. On the second bifurcation tree, the asymmetric period-5 motions exist for $\Omega \in (3.9425, 3.9625)$. The saddle-node bifurcation of $\Omega \approx 3.9625$ is for the onset of the asymmetric period-5 motion. At $\Omega \approx 3.9425$, the cascaded period-doubling bifurcation of the asymmetric period-5 motion introduces period-10, period-20... motions eventually to chaos.

The tree bifurcation trees for independent period-6 motions are presented in Figs. 4.6c,d for $\Omega \in (3.2, 4.4)$. Such period-6 motions are independent, which are not born from period-doubling bifurcations of period-3 motions. Two kinds of bifurcation trees of period-6 motion to chaos are observed. The first kind of bifurcation tree is from symmetric to asymmetric period-6 motions and to chaos, which is non-travelable. The second kind of bifurcation tree is directly from an asymmetric period-6 motion to chaos, which is travelable. There is only one bifurcation tree for independent asymmetric period-6 motions to chaos. Such an asymmetric period-6 motion is stable for $\Omega \in (3.7261, 3.7363)$, which are neither from saddle-node bifurcations of symmetric period-6 motion, nor from period-doubling bifurcations of asymmetric period-3 motions. Instead, such a bifurcation tree of the asymmetric period-6 motion to chaos exist independently. At $\Omega \approx 3.7363$, saddle-node bifurcation is for onset of the asymmetric period-6 motion. At $\Omega \approx 3.7261$, such an asymmetric period-6 motion is from stable to unstable through

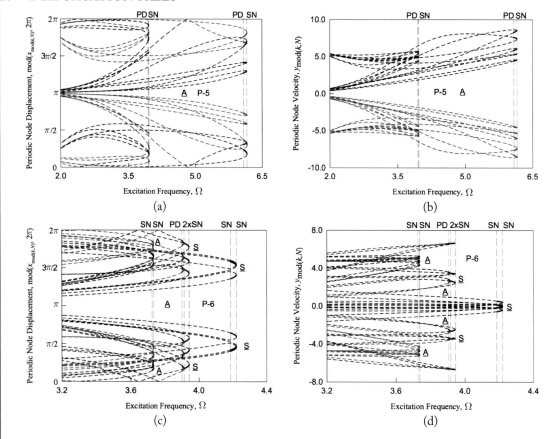

Figure 4.6: Period-5 motions to chaos with $\Omega \in (2.0, 6.5)$: (a) node displacement $x_{\text{mod}(k,N)}$, (b) node velocity $y_{\text{mod}(k,N)}$. Period-6 motions to chaos with $\Omega \in (3.2, 4.4)$: (c) node displacement $x_{\text{mod}(k,N)}$, (d) node velocity $y_{\text{mod}(k,N)}$. ($\alpha = 4.0$, $\delta = 0.1$, $Q_0 = 5.0$.)

the period-doubling bifurcation, which further introduces period-12, period-24... motions to chaos.

Two bifurcation trees for non-travelable symmetric period-6 motions are observed. The first one has a stable range of $\Omega \in (4.1857, 4.2197)$. The onset of such symmetric period-6 motion occurs at a saddle-node bifurcation of $\Omega \approx 4.2197$. The saddle-node bifurcation at $\Omega \approx 4.1857$ is for onset of asymmetric period-6 motion from the symmetric period-6 motion. The introduced asymmetric period-6 motions are not illustrated due to the very tiny stable ranges. The second branch of the symmetric period-6 motion is stable for the range $\Omega \in (3.9155, 3.9443)$. At $\Omega \approx 3.9443$, a saddle-node bifurcation is for the onset of the symmetric period-6 motion. At $\Omega \approx 3.9155$, a saddle-node bifurcation is for a pair of asymmetric period-6 motions from the symmetric period-6 motions. Such asymmetric motions are stable for $\Omega \in (3.9029, 3.9155)$. The cascaded period-doubling bifurcation of the asymmetric period-6 motion at $\Omega \approx 3.9029$ produces period-12, period-24... motions to chaos.

4.6   INDEPENDENT PERIOD-8, PERIOD-10, AND PERIOD-12 MOTIONS

More independent motions are observed in such a parametrically excited pendulum with excitation frequency. In Figs. 4.7, the bifurcation trees for independent period-8, period-10, and period-12 motions are illustrated. The bifurcation trees for independent period-8 motions are presented in Figs. 4.7a,b for $\Omega \in (3.0, 6.8)$. Two different bifurcation trees exist. The first bifurcation tree starts from a pair of independent asymmetric period-8 motions. Such asymmetric motions are neither from saddle-node bifurcations of symmetric period-8 motions nor from period-doubling bifurcations of asymmetric period-4 motions. The asymmetric period-8 motion is stable for $\Omega \in (6.6104, 6.5794)$. At $\Omega \approx 6.5794$, the saddle-node bifurcation is for the onset of the asymmetric period-8 motion. At $\Omega \approx 6.6104$, the period-doubling bifurcation of the asymmetric period-8 motion leads to the asymmetric period-8 motions, and the further period-doubling bifurcation is for period-16, period-32... motions to chaos. On the second independent bifurcation tree, the period-8 motions are symmetric. Such symmetric period-8 motion is not from period-doubling bifurcation of period-4 motion. Such symmetric period-8 motion is stable for $\Omega \in (3.3209, 3.3313)$. The saddle-node bifurcation at $\Omega \approx 3.3313$ is for the birth of the symmetric period-8 motion. At $\Omega \approx 3.3209$, a saddle-node bifurcation generates an asymmetric period-8 motion. Such an asymmetric period-8 motion is not illustrated due to a tiny stable range.

The bifurcation tree for a non-travelable, symmetric period-10 motion is presented in Figs. 4.7c,d. There is only one branch of bifurcation tree for such an independent period-10 motion, which is not from period-doubling of period-5 motions. Such a period-10 motion is stable for $\Omega \in (4.4781, 4.4939)$. At $\Omega \approx 4.4939$, a saddle-node bifurcation is for onset of the symmetric period-10 motion. At $\Omega \approx 4.4781$, the saddle-node bifurcation introduces a pair of asymmetric period-10 motions, which will not be presented herein.

The bifurcation trees for a non-travelable, symmetric period-12 motions are presented in Figs. 4.7e,f. There is one bifurcation tree of such an independent period-12 motion to chaos. Such a symmetric period-12 motion is not from period-doubling bifurcations of period-6 motions. The stable range for such a symmetric period-12 motion is $\Omega \in (6.4078, 6.4502)$. The onset of the symmetric period-12 motion occurs at a saddle-node bifurcation of $\Omega \approx 6.4502$. At $\Omega \approx 6.4078$, the saddle-node bifurcation introduces a pair of asymmetric period-12 motions. Once again, the introduced asymmetric period-12 motions are also not presented herein because of the very tiny stable range.

Similarly, bifurcation trees pertaining to other periodic motions can be determined in the parametric pendulum.

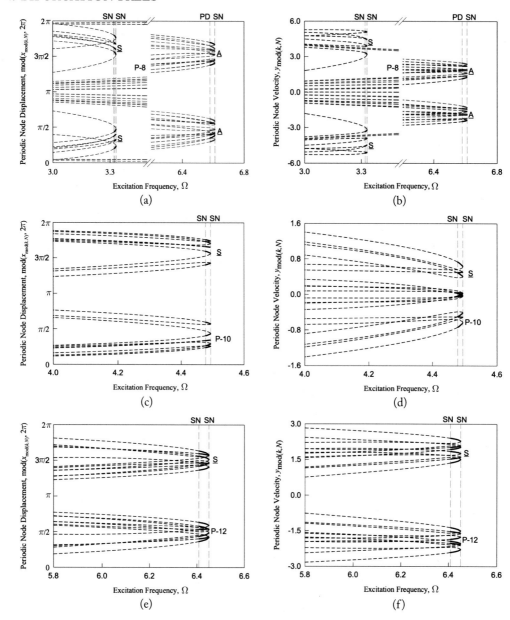

Figure 4.7: Bifurcation tree of period-8 motions to chaos varying with excitation frequency Ω. Period-8 motion: (a) node displacement $x_{\text{mod}(k,N)}$, (b) node velocity $y_{\text{mod}(k,N)}$. Period-10 motion: (a) node displacement $x_{\text{mod}(k,N)}$, (b) node velocity $y_{\text{mod}(k,N)}$. Period-12 motion: (a) node displacement $x_{\text{mod}(k,N)}$, (b) node velocity $y_{\text{mod}(k,N)}$. ($\alpha = 4.0$, $\delta = 0.1$, $Q_0 = 5.0$, $\Omega \in (3.0, 6.8)$.)

CHAPTER 5

Harmonic Frequency-Amplitude Characteristics

In this section, the harmonic frequency-amplitude characteristics are presented for period-1 static points, period-1, period-2, period-3, period-4 motions to chaos. In order to avoid abundant illustrations, other harmonic characteristics for period-5, period-6, period-8, period-10, and period-12 motions are not presented. The corresponding bifurcation trees are presented through harmonic frequency-amplitude curves of periodic node displacements $\text{mod}(x_{\text{mod}(k,N)}, 2\pi)$ for non-travelable periodic motions. For the travelable period-m motions, the harmonic analysis of periodic node velocities are presented. Because of $x_0 \neq x_{mT}$, the periodic node displacements cannot be used for the harmonic analysis of the periodic motions.

5.1 DISCRETE FOURIER SERIES

For period-m motions in the parametrically driven pendulum system, the discrete node points $\mathbf{x}_k^{(m)} = (x_k^{(m)}, y_k^{(m)})^T$ ($k = 0, 1, 2, \ldots, mN$) can be approximated through Fourier series, i.e.,

$$\mathbf{x}^{(m)}(t) \approx \mathbf{a}_0^{(m)} + \sum_{j=1}^{M} \mathbf{b}_{j/m} \cos(\frac{k}{m}\Omega t) + \mathbf{c}_{j/m} \sin(\frac{k}{m}\Omega t). \tag{5.1}$$

The $(2M + 1)$ unknown vector coefficients of $\mathbf{a}_0^{(m)}, \mathbf{b}_{j/m}, \mathbf{c}_{j/m}$ should be determined from the discrete nodes $\mathbf{x}_k^{(m)}$ ($k = 0, 1, 2, \ldots, mN$) with $mN + 1 \geq 2M + 1$. For $M = mN/2$, the node points $\mathbf{x}_k^{(m)}$ on the period-m motion can be expressed for $t_k \in [0, mT]$ as

$$\begin{aligned}
\mathbf{x}^{(m)}(t_k) \equiv \mathbf{x}_k^{(m)} &= \mathbf{a}_0^{(m)} + \sum_{j=1}^{mN/2} \mathbf{b}_{j/m} \cos(\frac{j}{m}\Omega t_k) + c_{j/m} \sin(\frac{j}{m}\Omega t_k) \\
&= \mathbf{a}_0^{(m)} + \sum_{j=1}^{mN/2} \mathbf{b}_{j/m} \cos(\frac{j}{m}\frac{2k\pi}{N}) + \mathbf{c}_{j/m} \sin(\frac{j}{m}\frac{2k\pi}{N}) \\
&\quad (k = 0, 1, \ldots, mN - 1),
\end{aligned} \tag{5.2}$$

where

$$T = \frac{2\pi}{\Omega} = N\Delta t; \quad \Omega t_k = \Omega k\Delta t = \frac{2k\pi}{N},$$

$$\mathbf{a}_0^{(m)} = \frac{1}{N}\sum_{k=0}^{mN-1}\mathbf{x}_k^{(m)},$$

$$\left.\begin{aligned}\mathbf{b}_{j/m} &= \frac{2}{mN}\sum_{k=0}^{mN-1}\mathbf{x}_k^{(m)}\cos(k\tfrac{2j\pi}{mN}),\\[2mm]\mathbf{c}_{j/m} &= \frac{2}{mN}\sum_{k=0}^{mN-1}\mathbf{x}_k^{(m)}\sin(k\tfrac{2j\pi}{mN})\end{aligned}\right\} \quad (j=1,2,\ldots,mN/2)$$

(5.3)

and

$$\mathbf{a}_0^{(m)} = (a_{01}^{(m)}, a_{02}^{(m)})^T, \quad \mathbf{b}_{j/m} = (b_{j/m1}, b_{j/m2})^T, \quad \mathbf{c}_{j/m} = (c_{j/m1}, c_{j/m2})^T. \quad (5.4)$$

The harmonic amplitudes and harmonic phases for the period-m motions are expressed by

$$\begin{aligned}A_{j/m1} &= \sqrt{b_{j/m1}^2 + c_{j/m1}^2}, & \phi_{j/m1} &= \arctan\frac{c_{j/m1}}{b_{j/m1}},\\[2mm]A_{j/m2} &= \sqrt{b_{j/m2}^2 + c_{j/m2}^2}, & \phi_{j/m2} &= \arctan\frac{c_{j/m2}}{b_{j/m2}}.\end{aligned} \quad (5.5)$$

Thus, the approximate expression of period-m motions in Eq. (5.1) becomes

$$\mathbf{x}^{(m)}(t) \approx \mathbf{a}_0^{(m)} + \sum_{j=1}^{mN/2}\mathbf{b}_{j/m}\cos(\frac{k}{m}\Omega t) + \mathbf{c}_{j/m}\sin(\frac{k}{m}\Omega t). \quad (5.6)$$

For the parametrically excited pendulum system, we have

$$\begin{Bmatrix} x^{(m)}(t) \\ y^{(m)}(t) \end{Bmatrix} \equiv \begin{Bmatrix} x_1^{(m)}(t) \\ x_2^{(m)}(t) \end{Bmatrix} \approx \begin{Bmatrix} a_{01}^{(m)} \\ a_{02}^{(m)} \end{Bmatrix} + \sum_{j=1}^{mN/2}\begin{Bmatrix} A_{j/m1}\cos(\frac{k}{m}\Omega t - \phi_{j/m1}) \\ A_{j/m2}\cos(\frac{k}{m}\Omega t - \phi_{j/m2}) \end{Bmatrix}. \quad (5.7)$$

To reduce illustrations, only harmonic amplitudes of displacement $x^{(m)}(t)$ for period-m motions will be presented for non-travelable period-m motions. However, the harmonic amplitudes for velocity $y^{(m)}(t)$ can also be achieved for travelable period-m motions in a similar fashion. Thus, the displacement for period-m motion is given by

$$x^{(m)}(t) \approx a_0^{(m)} + \sum_{j=1}^{mN/2}b_{j/m}\cos(\frac{k}{m}\Omega t) + c_{j/m}\sin(\frac{k}{m}\Omega t), \quad (5.8)$$

or

$$x^{(m)}(t) \approx a_0^{(m)} + \sum_{j=1}^{mN/2} A_{j/m} \cos(\frac{k}{m}\Omega t - \phi_{j/m}),$$ (5.9)

where

$$A_{j/m} = \sqrt{b_{j/m}^2 + c_{j/m}^2}, \qquad \phi_{j/m} = \arctan \frac{c_{j/m}}{b_{j/m}}.$$ (5.10)

In all plots of the harmonic frequency-amplitude curves, the black solid curves represent stable motions, while the red dashed curves are for unstable periodic motions. The acronyms "PD" and "SN" are for period-doubling and saddle-node bifurcation, respectively. "A" or "S" is for asymmetric or symmetric motion, accordingly. The travelable or non-travelable motions are marked by "Travelable" or "Non-travelable."

5.2   NON-TRAVELABLE PERIOD-1 STATIC POINTS TO CHAOS

The bifurcation trees of harmonic-amplitudes varying with excitation frequency are presented for the period-1 static points to chaos in Fig. 5.1. All period-m motions presented in Fig. 5.1 are non-travelable with $x_k = x_{k+mN}$ and $y_k = y_{k+mN}$. The harmonic frequency-amplitudes of displacements are carried out for the non-travelable period-1 static points to chaos. The constant term $a_0^{(m)}$ ($m = 1, 2$) is presented in Fig. 5.1(i) for the solution center at $\mathrm{mod}(a_0^{(m)}, l\pi) = 0$ ($l = 0, 1, 2, \ldots$). The bifurcation points can be clearly observed. As excitation frequency Ω changes, the period-doubling and saddle-node bifurcations of the period-1 static point are for symmetric period-2 motions and asymmetric period-1 motions, respectively. The pairs of asymmetric motions in $a_0^{(m)}$ is symmetric to each other around π. However, the harmonic amplitudes of $A_{k/2}$ ($k = 1, 2, 3, \ldots$) are of the same. The harmonic term $a_0^{(m)}$ for symmetric period-2 motions and period-1 static points is 2π or 0. All the asymmetric motions are from the symmetric motions through saddle-node bifurcations. In Fig. 5.1(ii), harmonic amplitude $A_{1/2}$ is presented. For period-1 motions, $A_{1/2} = 0$. The quantity level of period-2 motion is $A_{1/2} \sim 4.0$. In Fig. 5.1(iii), the primary harmonic amplitude A_1 is presented for both period-1 and period-2 motions. For symmetric period-2 motions and period-1 static points, $A_1 = 0$. For asymmetric motions, the quantity level is $A_1 \sim 3.0$. In Fig. 5.1(iv), harmonic amplitudes of $A_{3/2}$ are presented with a quantity level of $A_{3/2} \sim 2.1$. The other harmonic amplitudes $A_{k/2}$ ($k = 2l + 1,\ l = 2, 3, 4, \ldots$) will not be presented to avoid abundant illustrations. In Figs. 5.1(v)–5.1(x), the harmonic amplitudes A_k ($k = 2, 3, 4, 5, 6, 7$) are presented with different quantity levels. The maximum quantity levels are $A_2 \sim 1.2$, $A_3 \sim 1.0$, $A_4 \sim 0.4$, $A_5 \sim 0.12$ $A_6 \sim 0.06$, and $A_7 \sim 0.03$. The quantity level drops to 10^{-2} with harmonic order increase. On the other hand, the quantity level also drops as excitation frequency increases for each harmonic term. The quantity levels for $\Omega \in (2.0, 6.7)$ are $A_2 \sim 0.8$, $A_3 \sim 0.4$, $A_4 \sim 0.05$, $A_6 \sim 0.005$, and

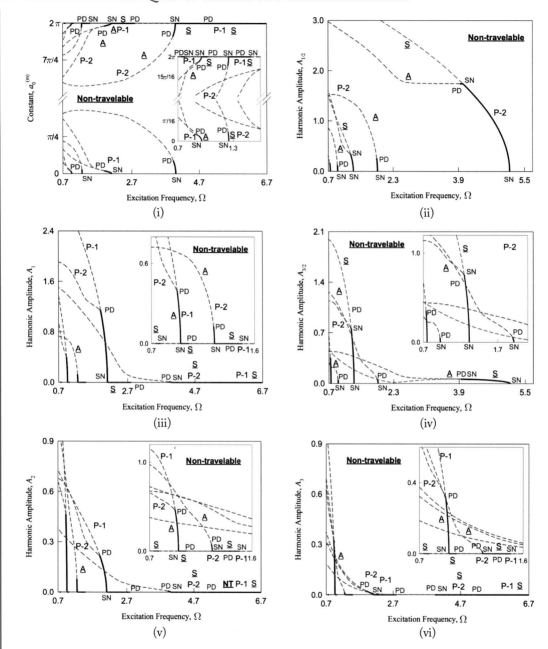

Figure 5.1: Harmonic frequency-amplitude characteristics for bifurcation trees of period-1 static points to chaos: (i) $a_0^{(m)}$ ($m = 1, 2$). (ii)–(vi) $A_{k/m}$ ($m = 2$, $k = 1, 2, 3, 4; 6, 8, \ldots, 28; 40, 41$); Parameters: ($\alpha = 4.0, \delta = 0.1, Q_0 = 5.0$, $\Omega \in (0.7, 6.7)$). (*Continues.*)

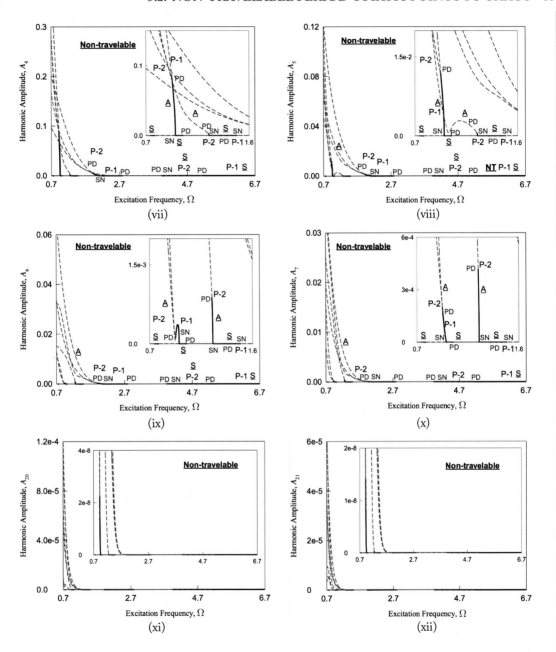

Figure 5.1: (*Continued.*) Harmonic frequency-amplitude characteristics for bifurcation trees of period-1 static points to chaos: (vii)–(xii) $A_{k/m}$ ($m = 2$, $k = 1, 2, 3, 4; 6, 8, \ldots, 28; 40, 41$); Parameters: ($\alpha = 4.0$, $\delta = 0.1$, $Q_0 = 5.0$, $\Omega \in (0.7, 6.7)$).

$A_9 \sim 10^{-3}$. In Figs. 5.1(xi) and (xii), harmonic amplitudes A_{20}, and A_{21} are presented. The overall quantity levels are $A_{20} \sim 10^{-4}$, $A_{21} \sim 10^{-4}$, and $A_{61} < 10^{-10}$.

5.3   TRAVELABLE PERIOD-1 MOTIONS TO CHAOS

Unlike period-1 static points to chaos, the bifurcation trees of period-1 motions to chaos are travelable, which means $x_k \neq x_{k+mN}$ and $y_k = y_{k+mN}$. The Fourier series of displacement for travelable period-1 motions to chaos cannot be carried out. Thus, the harmonic frequency-amplitudes of velocity will be computed, and the bifurcation trees of harmonic amplitudes are presented for the period-1 motions to chaos in Fig. 5.2.

The constant term $a_0^{(m)} = \Omega$ ($m = 1, 2, 4$) of velocity is presented in Fig. 5.2(i) for the positive branch of the asymmetric period-1 motion to chaos. For the negative branch of the asymmetric period-1 motion to chaos, $a_0^{(m)} = -\Omega$ is not presented. However, the other harmonic amplitudes are of the same for the positive and negative branches. In Fig. 5.2(ii), the harmonic amplitude $A_{1/4}$ is presented for period-4 motions. For period-1 and period-2 motions, $A_{1/4} = 0$. The overall quantity level is $A_{1/4} \sim 0.6$. In Fig. 5.2(iii), the harmonic amplitude of $A_{1/2}$ is presented for period-2 and period-4 motions. The period-doubling bifurcation from period-2 motions to period-4 motions can be clearly observed. The quantity level is $A_{1/2} \sim 2.5$. The harmonic amplitude of $A_{3/4}$ is presented in Fig. 5.2(iv) for period-4 motion only. Thus, $A_{3/4} = 0$ for period-1 and period-2 motions. The quantity level is $A_{3/4} \sim 1.0$. To avoid abundant illustrations, the harmonic amplitudes $A_{(2l-1)/4}$ ($l = 3, 4, \ldots$) will not be presented. In Fig. 5.2(v), the primary harmonic amplitude A_1 is presented for period-1 to period-4 motions. The asymmetric period-1 motions exist, which are independent of symmetric motions. Such an asymmetric period-1 motion has a saddle-node bifurcation for onset, and a period-doubling bifurcation for an asymmetric period-2 motion. Such an asymmetric period-2 motions develops asymmetric period-4 motions through a period-doubling bifurcation. For excitation frequency close to $\Omega = 1$, the quantity level is $A_1 \sim 4.2$. For high excitation frequency, the quantity level is $A_1 \sim 2.0$. In Fig. 5.2(vi), the harmonic amplitude $A_{3/2}$ is presented. For period-1 motions, $A_{3/2} = 0.0$, and the bifurcation tree from period-2 to period-4 motions can be clearly observed. The quantity level is $A_{3/2} \sim 2.0$. The harmonic amplitudes A_k ($k = 2, 3, \ldots, 5$) are then presented in Figs. 5.2(vii)–(x). For harmonic amplitudes $A_{2,3,\ldots,5}$, the quantity levels for lower excitation frequency ($\Omega \approx 1.0$) are quite different for higher excitation frequency. For excitation frequency close to $\Omega = 1$, the corresponding quantity levels are $A_2 \sim 2.1$, $A_3 \sim 1.5$, $A_4 \sim 0.5$, and $A_5 \sim 0.5$. For higher excitation frequency, the corresponding quantity levels are $A_2 \sim 0.5$, $A_3 \sim 0.3$, $A_4 \sim 0.02$ and $A_5 \sim 2 \times 10^{-3}$. For harmonic amplitudes A_k ($k = 20, 21$), the corresponding quantity levels are $A_{20} \sim 9 \times 10^{-5}$ and $A_{21} \sim 5 \times 10^{-5}$ for $\Omega \in (1.0, 1.3)$. However, $A_{20,21} \in (10^{-9}, 10^{-6})$ for $\Omega > 1.3$.

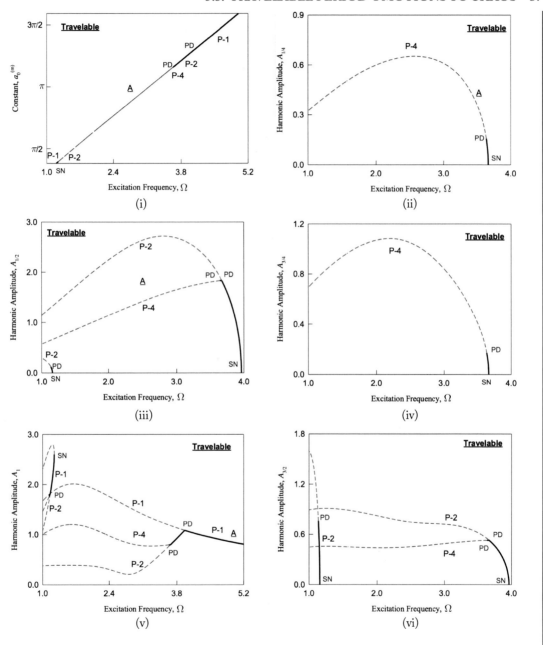

Figure 5.2: Harmonic frequency-amplitude characteristics for bifurcation trees of travelable period-1 motions to chaos based on velocity: (i) $a_0^{(m)}$ ($m = 1, 2, 4$). (ii)–(vi) $A_{k/m}$ ($m = 4$, $k = 1, 2, 3, 4; 6, 8, 12, \ldots, 24, 80, 84$); Parameters: ($\alpha = 4.0$, $\delta = 0.1$, $Q_0 = 5.0$, $\Omega \in (1.0, 5.2)$). (*Continues.*)

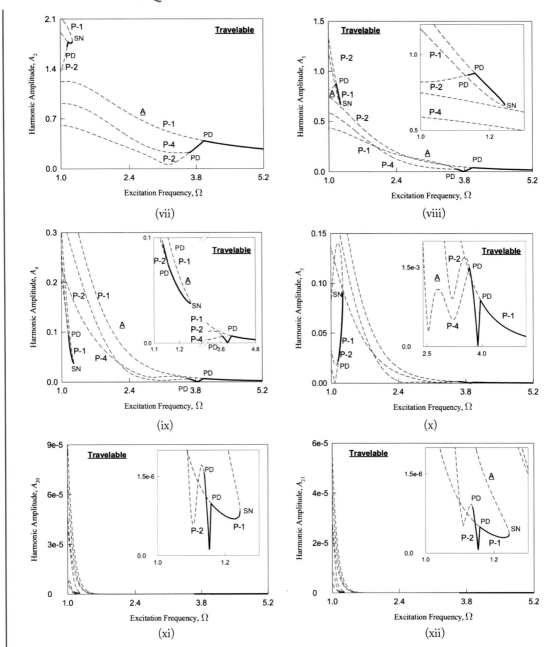

Figure 5.2: (*Continued.*) Harmonic frequency-amplitude characteristics for bifurcation trees of travelable period-1 motions to chaos based on velocity: (vii)–(xii) $A_{k/m}$ ($m = 4, k = 1, 2, 3, 4; 6, 8, 12, \ldots, 24, 80, 84$); Parameters: ($\alpha = 4.0$, $\delta = 0.1$, $Q_0 = 5.0$, $\Omega \in (1.0, 5.2)$).

5.4   NON-TRAVELABLE PERIOD-2 MOTIONS TO CHAOS

In Fig. 5.3, the bifurcation trees of period-2 motions to chaos are presented through frequency-amplitude curves. Such period-2 motions are independent, which are not from the period-doubling bifurcation of period-1 motions. All motions presented in Fig. 5.3 are non-travelable with $x_k = x_{k+mN}$ and $y_k = y_{k+mN}$. Thus, the harmonic-amplitude curves for the period-2 motion to chaos can be done through the corresponding displacements. The constant term $a_0^{(m)}$ ($m = 2$) is presented in Fig. 5.3(i) for the solution center at $\mathrm{mod}(a_0^{(m)}, l\pi) = 0$ ($l = 0, 1, 2, \ldots$). The three branches of bifurcation points can be clearly observed. For symmetric period-2 motions, $a_0^{(m)} = \pi$. All the paired asymmetric period-2 motions are from the symmetric period-2 motions. Such asymmetric period-2 motions are symmetric to each other at $a_0^{(m)} = \pi$. The asymmetric period-2 motions can develop asymmetric period-4, period-8 ... motions through period-doubling bifurcations. In Fig. 5.3(ii), the harmonic amplitude of $A_{1/2}$ is presented. Unlike the $a_0^{(m)}$ term, the saddle-node bifurcation for onset of the period-2 motion is observed for all three branches of bifurcation trees. The quantity level is $A_{1/2} \sim 5.0$. The primary harmonic amplitude A_1 is presented in Fig. 5.3(iii). For symmetric period-2 motions, $A_1 = 0.0$, and the asymmetric period-2 motions are introduced by saddle-node bifurcations. The quantity level is $A_1 \sim 1.5$. Similarly, the harmonic amplitudes of $A_{3/2}$, $A_{5/2}$, $A_{7/2}$, and $A_{9/2}$ are presented in Figs. 5.3(iv), (vi), and Figs. 5.3(viii), (x), respectively. The quantity level of $A_{3/2}$ is $A_{3/2} \sim 1.8$. As harmonic order increases, the quantity levels for higher excitation frequency Ω drops faster than lower frequency. For $\Omega < 1.0$, the quantity levels are $A_{5/2} \sim 1.5$, $A_{7/2} \sim 1.2$ and $A_{9/2} \sim 0.8$. However, for $\Omega > 1.0$, the quantity levels are $A_{5/2} \sim 0.5$, $A_{7/2} \sim 0.1$, and $A_{9/2} \sim 0.05$. In Figs. 5.3(v), 5.3(vii), (ix), the harmonic amplitudes of A_2, A_3, and A_4 are presented accordingly. The quantity level is $A_2 \sim 1.0$. With increasing the harmonic order, the quantity levels of the harmonic amplitudes for higher excitation frequency drop faster than for the lower frequency. For $\Omega < 1.0$, the quantity levels are $A_3 \sim 0.9$, and $A_4 \sim 0.8$. For $\Omega > 1.0$, the corresponding quantity levels are $A_3 \sim 0.1$ and $A_4 \sim 0.02$. Finally, to show the convergence of harmonic amplitudes, the harmonic amplitudes of A_{20} and A_{21} are presented in Fig. 5.3(xi) and (xii), respectively. The quantity levels are $A_{20,21} \sim 10^{-3}$ for $\Omega < 1.5$ and $A_{20,21} \sim 2 \times 10^{-9}$ for $\Omega > 1.5$.

5.5   PERIOD-3 MOTIONS

Period-3 motions can be non-travelable and travelable. The bifurcation tree for non-travelable period-3 motions to chaos is presented first, and then the travelable period-3 motions to chaos are presented.

In Fig. 5.4, the bifurcation trees of harmonic amplitudes are presented from non-travelable period-3 motions to chaos. The harmonic analysis of non-travelable period-3 motions to chaos is based on the corresponding displacement. The constant term $a_0^{(m)}$ ($m = 3$) is presented in Fig. 5.4(i). All motions in such bifurcation trees are asymmetric. For each bifurcation tree,

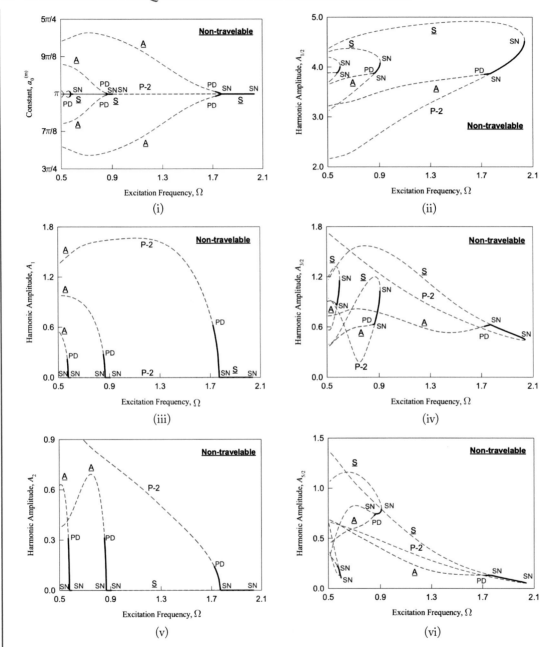

Figure 5.3: Harmonic frequency-amplitude characteristics for bifurcation trees of non-travelable period-2 motions to chaos based on displacement: (i) $a_0^{(m)}$ ($m = 2$). (ii)–(vi) $A_{k/m}$ ($m = 2$, $k = 1, 2, \cdots, 9, 40, 42$). Parameters: ($\alpha = 4.0$, $\delta = 0.1$, $Q_0 = 5.0$, $\Omega \in (0.5, 2.1)$). (*Continues.*)

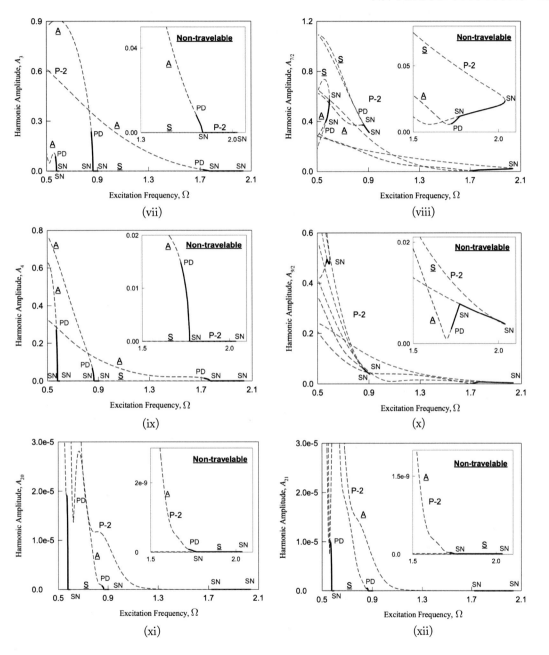

Figure 5.3: (*Continued.*) Harmonic frequency-amplitude characteristics for bifurcation trees of non-travelable period-2 motions to chaos based on displacement: (vii)–(xii) $A_{k/m}$ ($m = 2, k = 1, 2, \ldots, 9, 40, 42$). Parameters: ($\alpha = 4.0$, $\delta = 0.1$, $Q_0 = 5.0$, $\Omega \in (0.5, 2.1)$).

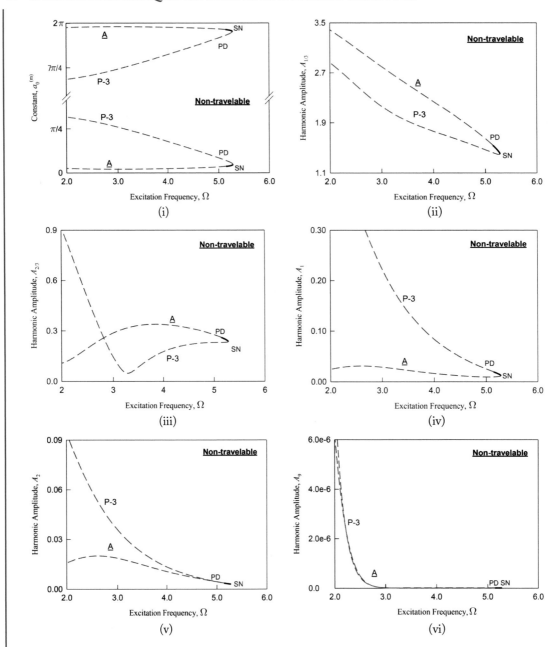

Figure 5.4: Harmonic frequency-amplitudes characteristics for bifurcation trees of non-travelable period-3 motion to chaos based on displacement: (i) $a_0^{(m)}$ ($m = 3$). (ii)–(vi) $A_{k/m}(m = 6, k = 1, 2, 3, 4; 6, 8; 27)$; Parameters: ($\alpha = 4.0$, $\delta = 0.1$, $Q_0 = 5.0$, $\Omega \in (2.0, 6.0)$).

there are two paired bifurcation trees, which are symmetric to π through the constant term of $a_0^{(m)}$. The bifurcation trees for the non-travelable period-3 motions are close to 0 or 2π. In Fig. 5.4(ii), the harmonic amplitude of $A_{1/3}$ is presented for the non-travelable period-3 motions. The corresponding quantity levels are $A_{1/3} \sim 3.5$. The harmonic amplitude $A_{2/3}$ is presented in Fig. 5.4(iii). The quantity levels of such harmonic amplitudes are $A_{2/3} \sim 0.8$. The primary harmonic amplitude A_1 is presented in Fig. 5.4(iv). The corresponding quantity level is $A_1 \sim 0.3$. The harmonic amplitudes of A_2 and A_9 are presented in Fig. 5.4(v) and (vi), respectively. The quantity levels for non-travelable periodic motions are $A_2 \sim 0.09$ and $A_9 \sim 10^{-5}$.

For travelable period-3 motion to chaos, the harmonic analysis is based on the corresponding velocity. The constant term $a_0^{(m)} = \Omega(m = 3, 6)$ of velocity is presented in Fig. 5.5(i) for the positive branch. For the negative branch of the asymmetric period-3 motion to chaos, $a_0^{(m)} = -\Omega$ is not presented. In Fig. 5.5(ii), the harmonic amplitude of $A_{1/6}$ is presented for period-6 motions only. For period-3 motions, $A_{1/6} = 0$. The quantity level is $A_{1/6} \sim 0.5$. In Fig. 5.5(iii), harmonic amplitude $A_{1/3}$ is presented for period-3 and period-6 motions. The period-doubling bifurcation from period-3 to period-6 motions can be clearly observed. Such an asymmetric period-3 motions has a saddle-node bifurcation for onset, and a period-doubling bifurcation for an asymmetric period-6 motion. The quantity level is $A_{1/3} \sim 3.0$. The harmonic amplitude of $A_{1/2}$ is presented in Fig. 5.5(iv) for period-6 motion only. Thus, $A_{1/2} = 0$ for period-3 motions, an the quantity level is $A_{1/2} \sim 1.2$. In Fig. 5.5(v), the harmonic amplitude of $A_{2/3}$ is presented for period-3 and period-6 motions with the quantity level of $A_{2/3} \sim 1.8$. The harmonic amplitudes $A_{5/6}$ is presented in Fig. 5.5(vi), which is also for period-6 motion only. Thus, $A_{5/6} = 0$ for period-3 motions, and the quantity level is $A_{5/6} \sim 0.8$. To avoid abundant illustrations, the harmonic amplitudes of $A_{(2l-1)/6}$ ($l = 4, 5, \ldots$) will not be presented herein. In Fig. 5.5(vii), the primary harmonic amplitude A_1 is presented for period-3 to period-6 motions. The quantity level is $A_1 \sim 1.0$. The harmonic amplitudes A_k ($k = 2, 3, \ldots, 5, 9$) are presented in Fig. 5.5(viii)–(xii). The corresponding quantity levels are $A_2 \sim 0.6$, $A_3 \sim 0.3$, $A_4 \sim 0.05$, and $A_5 \sim 0.02$. For the harmonic amplitude of A_9, the quantity level is $A_9 \sim 2.5 \times 10^{-4}$ for $\Omega \in (2.0, 3.0)$. However, $A_9 \in (10^{-8}, 10^{-6})$ for $\Omega > 3.0$.

5.6   PERIOD-4 MOTIONS

Period-4 motions also can be non-travelable and travelable, which are independent. Such period-4 motions are not from period-2 motion through the period-doubling bifurcation. The bifurcation treed for non-travelable period-4 motions to chaos are presented first through the displacement, and the travelable period-4 motions to chaos are presented through the velocity.

In Fig. 5.6, the bifurcation trees of harmonic amplitudes are presented from non-travelable period-3 motions to chaos through the displacement. The constant term $a_0^{(m)}$ ($m = 4$) is presented in Fig. 5.6(i). The two paired bifurcation trees are symmetric to π. The bifurcation trees for the non-travelable period-4 motions are close to 0 or 2π. In Fig. 5.6(ii), the harmonic am-

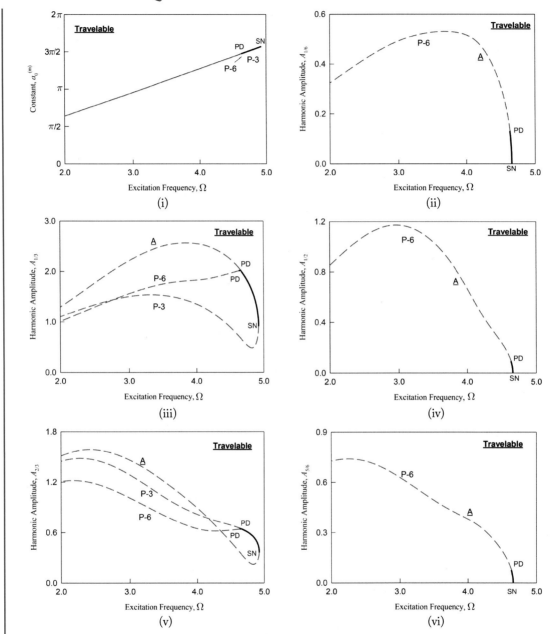

Figure 5.5: Harmonic frequency-amplitude characteristics for bifurcation trees of travelable period-3 motion to chaos based on velocity: (i) $a_0^{(m)}$ ($m = 3, 6$). (ii)–(vi) $A_{k/m}$ ($m = 6$, $k = 1, 2, \ldots, 6, 12, \ldots, 30, 54$); Parameters: ($\alpha = 4.0$, $\delta = 0.1$, $Q_0 = 5.0$, $\Omega \in (2.0, 6.0)$). (*Continues.*)

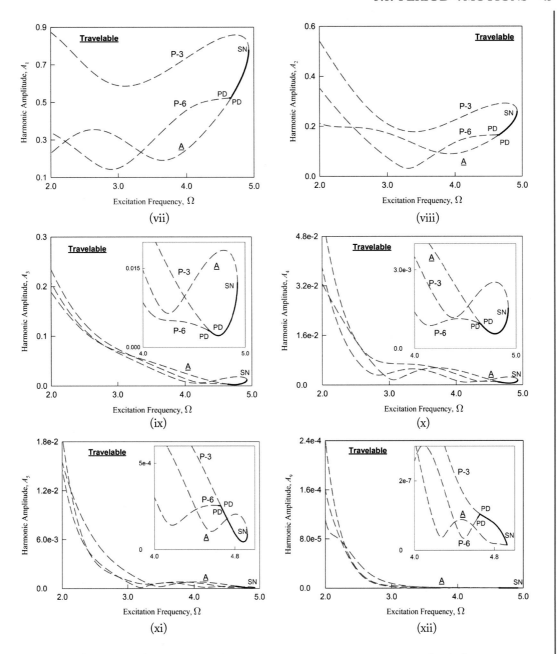

Figure 5.5: (*Continued.*) Harmonic frequency-amplitude characteristics for bifurcation trees of travelable period-3 motion to chaos based on velocity: (vii)–(xii) $A_{k/m}(m = 6,\ k = 1, 2, \ldots, 6, 12, \ldots, 30, 54)$; Parameters: ($\alpha = 4.0,\ \delta = 0.1,\ Q_0 = 5.0,\ \Omega \in (2.0, 6.0)$).

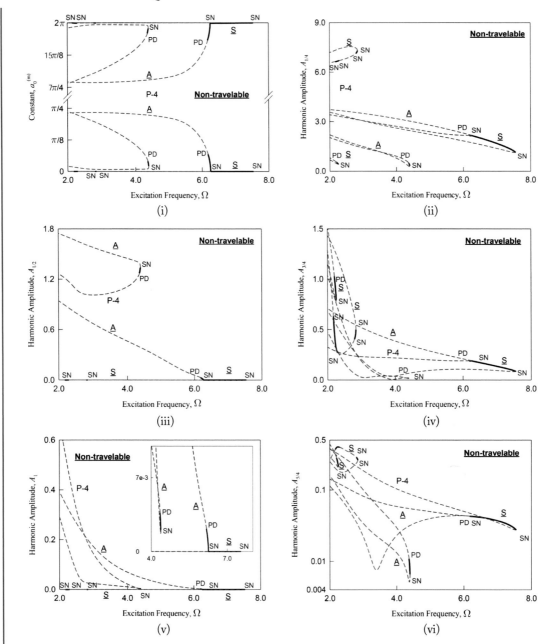

Figure 5.6: Harmonic frequency-amplitude characteristics for bifurcation trees of non-travelable period-4 motion to chaos based on displacement: (i) $a_0^{(m)}$ ($m = 3, 6$). (ii)–(vi) $A_{k/m}$ ($m = 6$, $k = 1, 2, \ldots, 8, 12, 16, 36$); Parameters: ($\alpha = 4.0$, $\delta = 0.1$, $Q_0 = 5.0$, $\Omega \in (2.0, 6.0)$). (*Continues.*)

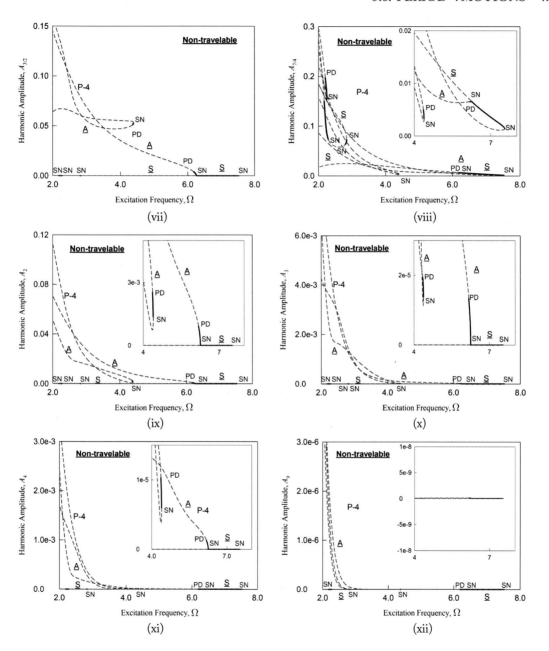

Figure 5.6: (*Continued.*) Harmonic frequency-amplitude characteristics for bifurcation trees of non-travelable period-4 motion to chaos based on displacement: (vii)–(xii) $A_{k/m}$ ($m = 6$, $k = 1, 2, \ldots, 8, 12, 16, 36$); Parameters: ($\alpha = 4.0$, $\delta = 0.1$, $Q_0 = 5.0$, $\Omega \in (2.0, 6.0)$).

plitude of $A_{1/4}$ is presented for the non-travelable period-4 motions. Three paired bifurcation trees for symmetric period-4 motions to chaos exists. Two paired asymmetric period-4 motions independently exist, which are not from the symmetric period-4 motions. The corresponding quantity levels are $A_{1/4} \sim 9.0$. The harmonic amplitude $A_{1/2}$ is presented in Fig. 5.6(iii). The quantity levels of such harmonic amplitudes are $A_{1/2} \sim 1.8$. In Fig. 5.6(iv), the harmonic amplitude of $A_{3/4}$ is presented for the non-travelable period-4 motions, which is similar to harmonic amplitude $A_{1/4}$. However, the corresponding quantity levels for $A_{1/4}$ and $A_{3/4}$ are different. The primary harmonic amplitude A_1 is presented in Fig. 5.6(v). The corresponding quantity level is $A_1 \sim 0.6$ for $\Omega \in (2.0, 4.0)$ and $A_1 \sim 0.01$ for $\Omega \in (4.0, 8.0)$. Similarly, the harmonic amplitudes of $A_{5/4}, A_{3/2}, A_{7/4}$ are presented in Figs. 5.6(vi)–(viii). The corresponding quantity levels are $A_{5/4} \sim 0.5$, $A_{3/2} \sim 0.15$, $A_{7/4} \sim 0.3$ for $\Omega \in (2.0, 4.0)$. The harmonic amplitudes of A_k ($k = 2, 3, 4, 9$) are presented in Figs. 5.6(ix)–(xii), respectively. The quantity levels for non-travelable period-4 motions are $A_2 \sim 0.12$, $A_3 \sim 0.01$, $A_4 \sim 3 \times 10^{-3}$, and $A_9 \sim 5 \times 10^{-6}$ for $\Omega \in (2.0, 4.0)$ and $A_2 \sim 10^{-2}$, $A_3 \sim 10^{-4}$, $A_4 \sim 5 \times 10^{-5}$, and $A_9 \sim 10^{-10}$ for $\Omega \in (4.0, 6.0)$.

For a travelable period-4 motion to chaos, the constant term $a_0^{(m)} = \Omega (m = 4)$ of velocity is presented in Fig. 5.7(i) for the positive branch. For the negative branch of the asymmetric period-4 motion to chaos, $a_0^{(m)} = -\Omega$ is not presented herein. In Fig. 5.7(ii), the harmonic amplitude of $A_{1/4}$ is presented for period-4 motions with the quantity level of $A_{1/4} \sim 3.0$. In Fig. 5.7(iii), harmonic amplitude $A_{1/2}$ is presented for period-4 motions with the quantity level of $A_{1/2} \sim 1.5$. In Fig. 5.7(iv), the $A_{5/4} \sim 0.7$, $A_{3/2} \sim 0.4$, $A_{7/4} \sim 0.6$ harmonic amplitude of $A_{3/4}$ is presented for period-4 motions, and the quantity level is $A_{3/4} \sim 1.5$. In Fig. 5.7(vii), the primary harmonic amplitude A_1 is presented for period-4 motions with the quantity level of $A_1 \sim 0.6$. $\Omega \in (4.0, 8.0)$. Similarly, the harmonic amplitudes of $A_{5/4}, A_{3/2}, A_{7/4}$ are presented in Figs. 5.7(vi)–(viii). The corresponding quantity levels are $A_{5/4} \sim 0.7$, $A_{3/2} \sim 0.4$, and $A_{7/4} \sim 0.6$. The harmonic amplitudes of A_k ($k = 2, 3, 4, 9$) are presented in Figs. 5.7(ix)–(xii), respectively. The quantity levels for travelable period-4 motions are $A_2 \sim 0.36$, $A_3 \sim 0.18$, $A_4 \sim 0.03$, $A_9 \sim 5 \times 10^{-5}$ for $\Omega \in (2.0, 4.0)$ and $A_2 \sim 0.2$, $A_3 \sim 0.02$, $A_4 \sim 4 \times 10^{-3}$, and $A_9 \sim 3 \times 10^{-8}$ for $\Omega \in (4.0, 6.0)$.

The harmonic frequency-amplitudes for independent period-5, period-6, period-8, period-10, and period-12 motions can be analyzed similarly.

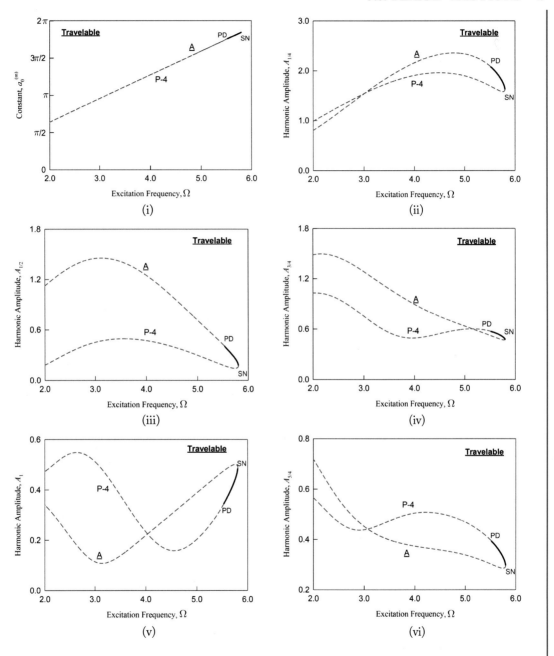

Figure 5.7: Harmonic frequency-amplitude characteristics for bifurcation trees of travelable period-4 motion to chaos based on velocity: (i) $a_0^{(m)}$ ($m = 3, 6$). (ii)–(vi) $A_{k/m}(m = 6, k = 1, 2, \ldots, 8, 12, 16, 36)$; Parameters: ($\alpha = 4.0$, $\delta = 0.1$, $Q_0 = 5.0$, $\Omega \in (2.0, 6.0)$). (*Continues.*)

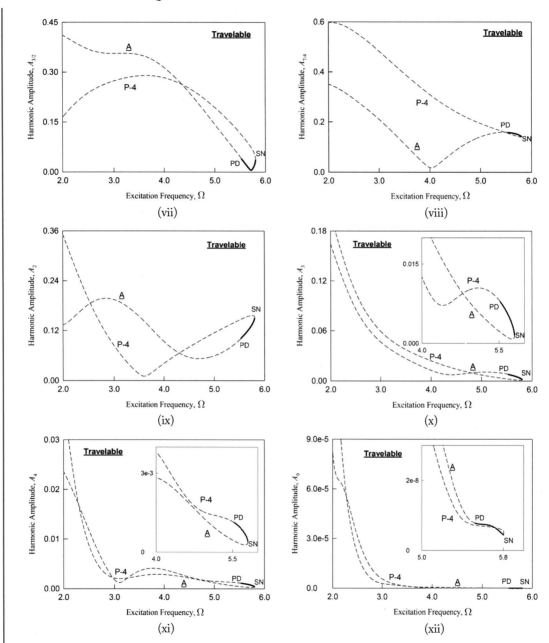

Figure 5.7: (*Continued.*) Harmonic frequency-amplitude characteristics for bifurcation trees of travelable period-4 motion to chaos based on velocity: (vii)–(xii) $A_{k/m}$ ($m = 6$, $k = 1, 2, \ldots, 8, 12, 16, 36$); Parameters: ($\alpha = 4.0$, $\delta = 0.1$, $Q_0 = 5.0$, $\Omega \in (2.0, 6.0)$).

CHAPTER 6

Non-Travelable Periodic Motions

In this chapter, various non-travelable periodic motions ($x_k = x_{k+mN}$, $y_k = y_{k+mN}$) will be illustrated with the same parameters in Eq. (4.1) for different excitation frequency. In all plots, the trajectories of periodic motions will be presented both numerically and analytically. To demonstrate harmonic effects on periodic motions, harmonic amplitudes and phases of periodic motions are presented. Numerical and analytical results will be presented by solid curves and hollow circular symbols, respectively. The initial condition and periodic nodes are indicated by green circles. For paired asymmetric motions, the black and red colors are for periodic motions on black and red branches of the bifurcation trees, respectively. The acronym "IC" is for initial condition. Finally, two different types of non-travelable periodic motions are observed: (i) librational motions and (ii) rotation motions with librations.

6.1 LIBRATIONAL PERIODIC MOTIONS

In this section, non-travelable librational periodic motions are presented. Period-1 to period-2 motions on the bifurcation tree of the period-1 static-points to chaos is presented first. The non-travelable librational period-3 motion are presented. Non-travelable librational symmetric period-4 to period-8 motions are presented.

6.1.1 PERIOD-1 TO PERIOD-2 MOTIONS

In Fig. 6.1, two paired librational asymmetric period-1 motions are illustrated for $\Omega = 2.0$, which are selected from the bifurcation trees of period-1 static-points to chaos. The initial conditions at $t_0 = 0.0$ are $x_0 \approx 0.0326$, $\dot{x}_0 \approx 2.4325$ for the black branch and $x_0 \approx 6.2506$, $\dot{x}_0 \approx -2.4325$ for the red branch. The displacement, trajectories, harmonic amplitudes, and phases are presented in Figs. 6.1a–d, respectively. Since the two paired asymmetric period-1 motions are non-travelable, the corresponding phase trajectories are two closed loops, as shown in Fig. 6.1b. Each trajectory goes through one cycle to complete a complete periodic orbit for one excitation period. The two asymmetric trajectories are skew symmetric to each other about the point $(\pi, 0)$. To understand the nonlinearity of such period-1 motions, the harmonic amplitude spectrum of displacement for period-1 motions are presented in Fig. 6.1c. The pair of asymmetric motions possess the exact same harmonic amplitudes except for the constant term

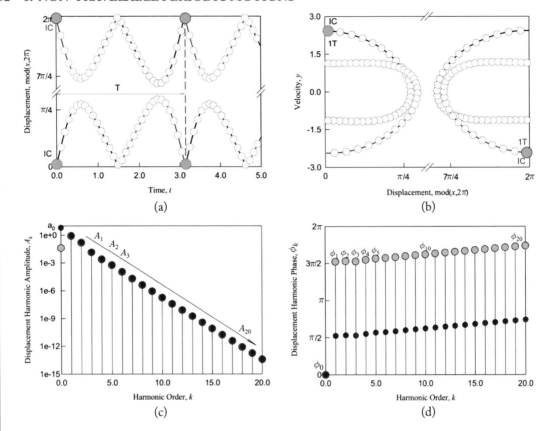

Figure 6.1: Non-travelable swinging asymmetric period-1 motion ($\Omega = 2.0$): (a) displacement, (b) trajectories, (c) harmonic amplitudes, and (d) harmonic phases. IC: $t_0 = 0.0$, $x_0 \approx$ 0.0326, $\dot{x}_0 \approx 2.4325$ for black branch and $x_0 \approx 6.2506$, $\dot{x}_0 \approx -2.4325$ for red branch. ($\alpha = 4.0$, $\delta = 0.1$, $Q_0 = 5.0$.)

a_0. For constant terms, the difference between the red and black branches are $a_0^r - a_0^b = 2\pi$. For primary harmonic amplitudes, the most significant amplitudes are $A_1 \approx 0.8492$, $A_2 \approx 0.1581$, $A_3 \approx 0.0137$, $A_4 \approx 2.3816 \times 10^{-3}$, and $A_5 \approx 5.6402 \times 10^{-4}$. Overall, the harmonic amplitudes converge quickly to $A_k < 10^{-12}$ ($k \geq 20$). The corresponding harmonic phases are presented in Fig. 6.1d with $\phi_k^b = \mathrm{mod}(\phi_k^r + \pi, 2\pi)$.

After a period-doubling bifurcation of the asymmetric period-1 motion, an asymmetric period-2 motion is developed from such an asymmetric period-1 motion. The paired asymmetric period-2 motions are presented in Fig. 6.2 for $\Omega = 1.90$. Such period-2 motions are still non-travelable. The initial conditions are $x_0 \approx 0.0166$, $\dot{x}_0 \approx 3.5889$ for the black branch and $x_0 \approx 6.2666$, $\dot{x}_0 \approx -3.5889$ for the red branch. The phase trajectories of the asymmetric period-

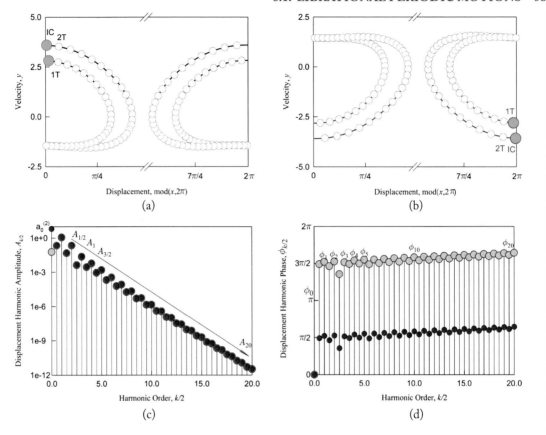

Figure 6.2: Non-travelable swinging asymmetric period-2 motion at $\Omega = 1.90$: (a) trajectory ($x_0 \approx 0.0166$, $\dot{x}_0 \approx 3.5889$ black), (b) trajectory ($x_0 \approx 6.2666$, $\dot{x}_0 \approx -3.5889$, red), (c) harmonic amplitudes, and (d) harmonic phases. ($\alpha = 4.0$, $\delta = 0.1$, $Q_0 = 5.0$.)

2 motions on the black and red branches are presented in Figs. 6.2a and b, respectively. The two asymmetric period-2 motions have two cycles for two periods. Symmetry of the two paired phase trajectories are skew symmetric to the point $(\pi, 0)$. The harmonic amplitudes for the two paired asymmetric period-2 motions are presented in Fig. 6.2c. For the period-2 motions, the harmonic amplitudes of $A_{1/2}$, $A_{3/2}$, and $A_{5/2} \dots$ can be observed. The constant terms have $a_0^{(2)r} - a_0^{(2)b} = 2\pi$, while all harmonic amplitudes are of the same for the period-2 motion on the black and red branches. The main primary harmonic amplitudes are $A_1 \approx 1.1505$, $A_2 \approx 0.2178$, $A_3 \approx 0.0229$, $A_4 \approx 5.9022 \times 10^{-3}$, and $A_5 \approx 1.5609 \times 10^{-3}$. The main subharmonic amplitudes are $A_{1/2} \approx 0.2180$, $A_{3/2} \approx 0.0468$, $A_{5/2} \approx 4.1945 \times 10^{-3}$, $A_{7/2} \approx 2.8201 \times 10^{-3}$, and $A_{9/2} \approx 9.0352 \times 10^{-4}$. The harmonic amplitudes converge to about $A_{20} \sim 10^{-12}$. The corre-

sponding harmonic phases for the two paired asymmetric period-2 motions are presented in Fig. 6.2d with $\phi_{k/2}^b = \mathrm{mod}(\phi_{k/2}^r + \pi, 2\pi)$.

6.1.2 ASYMMETRIC PERIOD-3 MOTIONS

In Fig. 6.3, two paired asymmetric period-3 motions are presented for $\Omega = 5.20$. The initial conditions are $x_0 \approx 1.3207$, $\dot{x}_0 \approx -1.2472$ for the black branch and $x_0 \approx 4.9625$, $\dot{x}_0 \approx 1.2472$ for the red branch. The phase trajectories of the asymmetric period-3 motions on the black and red branches are presented in Fig. 6.3a and b, respectively. Such asymmetric period-3 motions are non-travelable. The phase trajectories are closed loops in phase plane. Each trajectory for the asymmetric period-3 motion has one cycle in phase plane for three excitation periods. The tra-

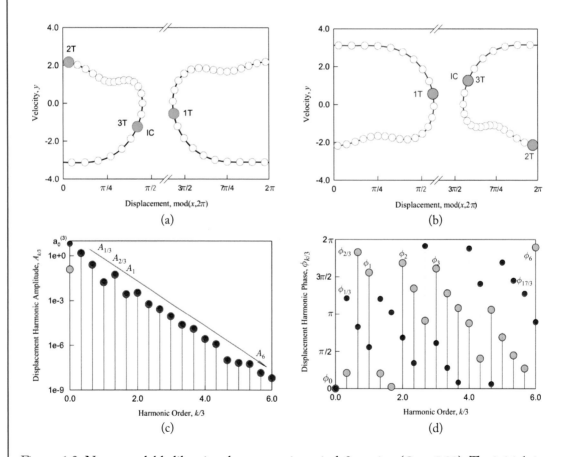

Figure 6.3: Non-travelable librational asymmetric period-3 motion ($\Omega = 5.20$). The initial time $t_0 = 0.0$: (a) trajectory ($x_0 \approx 1.3207$, $\dot{x}_0 \approx -1.2472$, black), (b) trajectory ($x_0 \approx 4.9625$, $\dot{x}_0 \approx 1.2472$, red), (c) harmonic amplitudes, and (d) harmonic phases. ($\alpha = 4.0$, $\delta = 0.1$, $Q_0 = 5.0$.)

jectories of the black and red branches are skew symmetric to each other about the point $(\pi, 0)$. The corresponding harmonic amplitudes of displacement are presented in Fig. 6.3c. For such asymmetric period-3 motions, the third-order terms can be observed. The paired asymmetric motions possess the exact same harmonic amplitudes. The two constant terms of $a_0^{(3)}$ possess a difference of 2π between the red and black branches (i.e., $a_0^{(3)r} - a_0^{(3)b} = 2\pi$). The main primary amplitudes are $A_1 \approx 0.0170$, $A_2 \approx 3.3360 \times 10^{-3}$, $A_3 \approx 8.9761 \times 10^{-5}$, $A_4 \approx 2.6519 \times 10^{-6}$, and $A_5 \approx 6.6904 \times 10^{-8}$. The main subharmonic amplitudes are $A_{1/3} \approx 1.4934$, $A_{2/3} \approx 0.2523$, $A_{4/3} \approx 0.0540$, $A_{5/3} \approx 2.6845 \times 10^{-3}$, $A_{7/3} \approx 5.7808 \times 10^{-4}$, and $A_{8/3} \approx 2.6592 \times 10^{-4}$. The subharmonic amplitudes possess quantity levels higher than the adjacent primary amplitudes. Overall, the harmonic amplitudes of $A_{k/3}(k = 1, 2, \ldots)$ converge to $A_{10} \sim 10^{-10}$. The corresponding harmonic phases are presented in Fig. 6.3d with $\phi_{k/3}^b = \mathrm{mod}(\phi_{k/3}^r + \pi, 2\pi)$. The harmonic phases for constant are zero because the two constants are positive.

6.1.3   SYMMETRIC AND ASYMMETRIC PERIOD-4 MOTIONS

From the bifurcation trees for independent period-4 motions to chaos, the non-travelable, librational symmetric period-4 to asymmetric period-8 motions are discussed. In Fig. 6.4, a symmetric librational period-4 motion is presented for $\Omega = 7.0$. The initial condition is $x_0 \approx 5.1389$ and $\dot{x}_0 \approx 2.0377$. The time-history of displacement for such a symmetric period-4 motion is presented in Fig. 6.4a, and the slow movement is observed. The phase trajectory of such a symmetric period-4 motion is illustrated in Fig. 6.4b. The trajectory of such a non-travelable period-4 motion forms a closed loop. Such a trajectory is symmetric to the point $(\pi, 0)$. The corresponding harmonic amplitudes are presented in Fig. 6.4c. For the constant term, $a_0^{(4)} \approx 6.28318$ and $\mathrm{mod}(a_0^{(4)}, 2\pi) \approx 0$. For the symmetric motions, the constant term can be $A_{0/4} = 0.0$. For such a symmetric period-4 motion, $A_{2l/4} = 0.0$, but $A_{(2l-1)/4} \neq 0.0$, $(l = 1, 2, 3 \ldots)$. The main harmonic amplitudes are $A_{1/4} \approx 1.6705$, $A_{3/4} \approx 0.1350$, $A_{5/4} \approx 0.0371$, $A_{7/4} \approx 3.5746 \times 10^{-3}$, and $A_{9/4} \approx 2.9680 \times 10^{-4}$. For $k = 31$, $A_{31/4} \sim 10^{-12}$. The harmonic phases for such a symmetric period-4 motion are presented in Fig. 6.4d. The harmonic phase distribution is for $\phi_{k/4} \in (0, 2\pi)$.

At $\Omega = 6.20$, the paired asymmetric period-4 motions are illustrated in Fig. 6.5. Such asymmetric period-4 motions are non-travelable. The initial conditions are $x_0 \approx 1.7598$, $\dot{x}_0 \approx -2.0475$ for the black branch, and $x_0 \approx 4.5234$, $\dot{x}_0 \approx 2.0475$ for the red branch. The trajectories of such coexisting asymmetric motions are presented in Figs. 6.5a and b, respectively. The coexisting paired trajectories of asymmetric period-4 motions on the black and red branches are skew symmetric to the point $(\pi, 0)$. The corresponding harmonic amplitudes and phases are presented in Figs. 6.5c and d, respectively. For constant term, $a_0^{(4)b} \approx 0.1820$ and $a_0^{(4)r} \approx 6.1012 = \mathrm{mod}(-0.1820, 2\pi)$ for the black and red branches, respectively. The main primary harmonic amplitudes of A_k $(k = 1, 2, \ldots)$ are $A_1 \approx 1.8104 \times 10^{-3}$, $A_2 \approx 7.8068 \times 10^{-4}$, $A_3 \approx 1.1384 \times 10^{-5}$, $A_4 \approx 1.1028 \times 10^{-6}$, and $A_5 \approx 2.9672 \times 10^{-8}$. The main subharmonic amplitudes of $A_{(2l-1)/2}$ $(l = 1, 2, \ldots)$ are $A_{1/2} \approx 0.0236$,

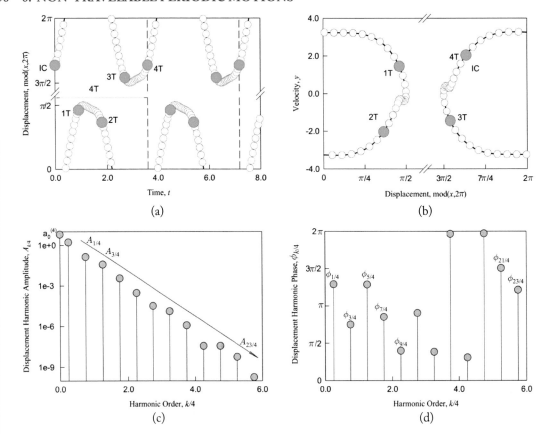

Figure 6.4: Non-travelable librational symmetric period-4 motion ($\Omega = 7.00$): (a) displacement, (b) trajectory, (c) harmonic amplitudes, and (d) harmonic phases. ($t_0 = 0.0$ $x_0 \approx 5.1389$, $\dot{x}_0 \approx 2.0377$), ($\alpha = 4.0$, $\delta = 0.1$, $Q_0 = 5.0$).

$A_{3/2} \approx 3.3965 \times 10^{-3}$, $A_{5/2} \approx 6.9618 \times 10^{-5}$, $A_{7/2} \approx 6.2216 \times 10^{-6}$, and $A_{9/2} \approx 7.6908 \times 10^{-8}$. Finally, the main subharmonic amplitudes of $A_{(2l-1)/4}$ ($l = 1, 2, \ldots$) are $A_{1/4} \approx 2.1488$, $A_{3/4} \approx 0.1887$, $A_{5/4} \approx 0.0429$, $A_{7/4} \approx 6.5392 \times 10^{-3}$, $A_{9/4} \approx 1.0881 \times 10^{-3}$, $A_{11/4} \approx 9.6651 \times 10^{-6}$, $A_{13/4} \approx 2.7466 \times 10^{-5}$, $A_{15/4} \approx 5.8305 \times 10^{-6}$, $A_{17/4} \approx 6.0938 \times 10^{-7}$, and $A_{19/4} \approx 4.2086 \times 10^{-8}$. Overall, the quantity level of the harmonic amplitudes reduces very quickly to $A_5 \sim 10^{-8}$. In Fig. 6.5d, the harmonic phases for constant harmonic terms both equal zero for the black and red branches. On the other hand, for all other harmonic phases, there is a difference of π between the black and red branches (i.e., $\phi_{k/4}^b = \mathrm{mod}(\phi_{k/4}^r + \pi, 2\pi)$) for $k = 1, 2, 3, \ldots$).

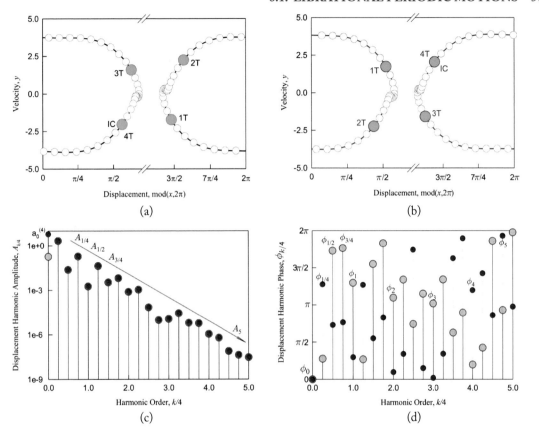

Figure 6.5: Non-travelable librational asymmetric period-4 motion ($\Omega = 6.20$): (a) trajectory ($x_0 \approx 1.7598$, $\dot{x}_0 \approx -2.0475$, black), (b) trajectory ($x_0 \approx 4.5234$, $\dot{x}_0 \approx 2.0475$, red), (c) harmonic amplitudes, and (d) harmonic phases. ($\alpha = 4.0$, $\delta = 0.1$, $Q_0 = 5.0$.)

6.1.4 INDEPENDENT PERIOD-8, PERIOD-10, AND PERIOD-12 MOTIONS

Two paired asymmetric librational period-8 motions are presented in Fig. 6.6 for $\Omega = 6.60$. Such period-8 motions are non-travelable, which are independent of period-4 motions. The corresponding trajectories of the black and red branches are illustrated in Figs. 6.6a and b, respectively. The initial conditions are $x_0 \approx 1.2845$, $\dot{x}_0 \approx 1.9033$ and $x_0 \approx 4.9987$, $\dot{x}_0 \approx -1.9033$ on the black and red branches, respectively. The skew symmetry of the two paired phase trajectories about the point $(\pi, 0)$ is observed. Such phase trajectory for asymmetric period-8 possesses two cycles through eight excitation periods. The harmonic amplitudes and phases for such paired asymmetric period-8 motions are presented in Figs. 6.6c and d, respectively. The two asymmetric period-8 motions possess the same harmonic amplitudes except for the con-

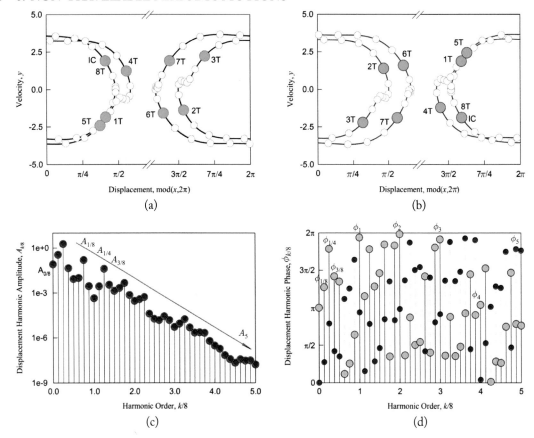

Figure 6.6: Non-travelable librational asymmetric period-8 motion ($\Omega = 6.60$): (a) trajectory ($x_0 \approx 1.2845$, $\dot{x}_0 \approx 1.9033$, black), (b) trajectory ($x_0 \approx 4.9987$, $\dot{x}_0 \approx -1.9033$, red), (c) harmonic amplitudes, and (d) harmonic phases. ($\alpha = 4.0, \delta = 0.1, Q_0 = 5.0$.)

stant terms. The two constants are $A_{0/8} = -a_0^{(8)b} \approx 0.0799$ and $A_{0/8} = a_0^{(8)r} \approx 0.0799$ for the black and red branches, respectively. The main primary harmonic amplitudes of $A_{8l/8} = A_l$ ($l = 1, 2, \ldots$) are $A_1 \approx 4.3970 \times 10^{-4}$, $A_2 \approx 2.9059 \times 10^{-4}$, and $A_3 \approx 5.1203 \times 10^{-6}$. The subharmonic terms of $A_{4(2l-1)/8} = A_{(2l-1)/2}$ ($l = 1, 2, \ldots$) also possess $A_{1/2} \approx 8.4174 \times 10^{-3}$, $A_{3/2} \approx 1.3675 \times 10^{-3}$, $A_{5/2} \approx 1.8247 \times 10^{-5}$ .... However, the main subharmonic amplitudes of $A_{2(2l-1)/8} = A_{(2l-1)/4}$ ($l = 1, 2, \ldots$) are $A_{1/4} \approx 1.8654$, $A_{3/4} \approx 0.1546$, $A_{5/4} \approx 0.0402$, $A_{7/4} \approx 4.5083 \times 10^{-3}$, $A_{9/4} \approx 5.2117 \times 10^{-4}$, $A_{11/4} \approx 2.7114 \times 10^{-5}$. For period-8 motions, the subharmonic amplitudes of $A_{(2l-1)/8}$ ($l = 1, 2, \ldots$) are $A_{1/8} \approx 0.3485$, $A_{3/8} \approx 0.0443$, $A_{5/8} \approx 9.8970 \times 10^{-3}$, $A_{7/8} \approx 2.8277 \times 10^{-3}$, $A_{9/8} \approx 2.7546 \times 10^{-3}$, $A_{11/8} \approx 3.4768 \times 10^{-3}$, $A_{13/8} \approx 2.0926 \times 10^{-3}$, $A_{15/8} \approx 7.1833 \times 10^{-4}$, $A_{17/8} \approx 3.7660 \times 10^{-4}$, $A_{19/8} \approx 4.0692 \times$

10^{-5}, $A_{21/8} \approx 1.4892 \times 10^{-5}$, and $A_{23/8} \approx 1.4883 \times 10^{-5}$. The harmonic phases in Fig. 6.6d give $\phi^b_{k/8} = \mathrm{mod}(\phi^r_{k/8} + \pi, 2\pi)$ $(k = 1, 2, 3, \ldots)$. The harmonic phases for constant terms are $\phi^{(8)b}_0 = \pi$ and $\phi^{(8)r}_0 = 0$.

Finally, non-travelable librational symmetric period-10 and period-12 motions are illustrated. The phase trajectory and harmonic amplitudes of a symmetric period-10 motion is illustrated in Figs. 6.7a and b for $\Omega = 4.49$. In a similar fashion, the phase trajectory and harmonic amplitudes of a symmetric period-12 motion are shown in Figs. 6.7c and d for $\Omega = 6.43$. The initial conditions are $x_0 \approx 1.8245$, $\dot{x}_0 \approx -0.0376$, and $x_0 \approx 1.7708$, $\dot{x}_0 \approx -1.9692$ for the period-10 and period-12 motions, respectively. Both of period-10 and period-12 motions are symmetric to the point $(\pi, 0)$.

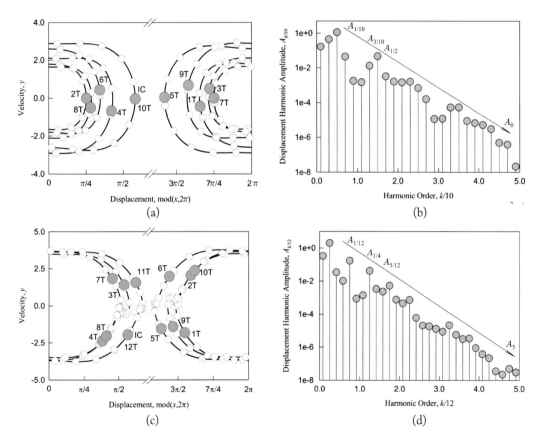

Figure 6.7: Non-travelable symmetric period-10 motion ($\Omega = 4.49$, $x_0 \approx 1.8245$, $\dot{x}_0 \approx -0.0376$): (a) trajectory and (b) harmonic amplitudes. Non-travelable symmetric period-12 motion ($\Omega = 6.43$, $x_0 \approx 1.7708$, $\dot{x}_0 \approx -1.9692$): (c) trajectory and (d) harmonic amplitudes. ($\alpha = 4.0$, $\delta = 0.1$, $Q_0 = 5.0$.)

The trajectory of the period-10 motion is demonstrated in Fig. 6.7a. Such trajectory has five cycles to form a close orbit with ten excitation periods. The corresponding harmonic amplitudes for the asymmetric period-10 motions are presented in Fig. 6.7b. $A_{0/10} = a_0^{(10)} = 0$ and $A_{(2l)/10} = A_{l/5} = 0$ $(l = 1, 2, \ldots)$. However, $A_{(2l-1)/10} \neq 0$ $(l = 1, 2, \ldots)$. The main subharmonic amplitudes are $A_{1/10} \approx 0.1666$, $A_{3/10} \approx 0.4491$, $A_{1/2} \approx 1.1339$, $A_{7/10} \approx 0.0436$, $A_{9/10} \approx 1.7910 \times 10^{-3}$, $A_{11/10} \approx 1.4832 \times 10^{-3}$, $A_{13/10} \approx 0.0131$, $A_{3/2} \approx 0.0466$, $A_{17/10} \approx 3.2480 \times 10^{-3}$, $A_{19/10} \approx 1.6102 \times 10^{-3}$, $A_{21/10} \approx 1.4659 \times 10^{-3}$, $A_{23/10} \approx 1.5898 \times 10^{-3}$, $A_{5/2} \approx 6.9232 \times 10^{-4}$, $A_{27/10} \approx 1.5927 \times 10^{-4}$, and $A_{29/10} \approx 1.1191 \times 10^{-5}$. For 25 subharmonic terms, we have $A_{49/10} \sim 10^{-8}$. Thus, for such symmetric period-10 motion, 25 subharmonic terms should be used.

The phase trajectory of the period-12 motion is presented in Fig. 6.7c. Such symmetric trajectory goes through three cycles to complete a periodic orbit for 12 excitation periods. The corresponding harmonic amplitudes of such symmetric period-12 motion is illustrated in Fig. 6.7d. $A_{0/12} = a_0^{(12)} = 0$ and $A_{(2l)/12} = A_{l/6} = 0$ $(l = 1, 2, \ldots)$. However, $A_{(2l-1)/12} \neq 0$ $(l = 1, 2, \ldots)$. The main subharmonic terms of $A_{(2l-1)/12}$ are $A_{1/12} \approx 0.3292$, $A_{1/4} \approx 1.9910$, $A_{5/12} \approx 0.0340$, $A_{7/12} \approx 0.0102$, $A_{3/4} \approx 0.1691$, $A_{11/12} \approx 8.7164 \times 10^{-4}$, $A_{13/12} \approx 1.3971 \times 10^{-3}$, $A_{5/4} \approx 0.0416$, $A_{17/12} \approx 3.3573 \times 10^{-3}$, $A_{19/12} \approx 2.3210 \times 10^{-3}$, $A_{7/4} \approx 5.3416 \times 10^{-3}$, $A_{23/12} \approx 7.6223 \times 10^{-4}$, $A_{25/12} \approx 4.5669 \times 10^{-4}$, $A_{9/4} \approx 7.3453 \times 10^{-4}$, and $A_{29/12} \approx 6.0345 \times 10^{-5}$. For 30 subharmonic terms, we have $A_{59/12} \sim 10^{-8}$. Thus, for such symmetric period-12 motion, 30 subharmonic terms should be used.

6.2 ROTATIONAL PERIODIC MOTIONS

In this section, non-travelable rotational periodic motions are illustrated from the bifurcation trees presented previously.

6.2.1 SYMMETRIC AND ASYMMETRIC PERIOD-6 MOTIONS

From the bifurcation trees of independent period-6 motions, a symmetric rotational period-6 motion is demonstrated in Fig. 6.8 for $\Omega = 3.93$. The initial condition is $x_0 \approx 1.6855$, and $\dot{x}_0 \approx -2.5981$. The displacement response of the symmetric period-6 motion is presented in Fig. 6.8a. For six excitation periods, the symmetric periodic motion returns back to the initial condition. The trajectory of such symmetric period-6 motion possesses a libration with two full clockwise rotations and then two full counter-clockwise rotations to complete a closed orbit, as shown in Fig. 6.8b. Such a trajectory is symmetric to the point $(\pi, 0)$. The corresponding harmonic amplitudes of such a symmetric period-6 motion is presented in Fig. 6.8c. The constant harmonic term possesses an amplitude of $a_0^{(6)} \approx 2\pi$ and $\mathrm{mod}(a_0^{(6)}, 2\pi) \approx 0$. $A_{2l/6} = A_{l/3} = 0$ but $A_{(2l-1)/6} \neq 0$ $(l = 1, 2, 3 \ldots)$. The main subharmonic amplitudes of $A_{(2l-1)/6}$ are $A_{1/6} \approx 8.6248$, $A_{1/2} \approx 0.0248$, $A_{5/6} \approx 0.1017$, $A_{7/6} \approx 0.1970$, $A_{3/2} \approx 0.0193$, $A_{11/6} \approx 7.4647 \times 10^{-3}$, $A_{13/6} \approx 0.0249$, $A_{5/2} \approx 0.0126$, $A_{17/6} \approx 1.5580 \times 10^{-3}$, $A_{19/6} \approx 8.3301 \times 10^{-4}$, $A_{7/2} \approx$

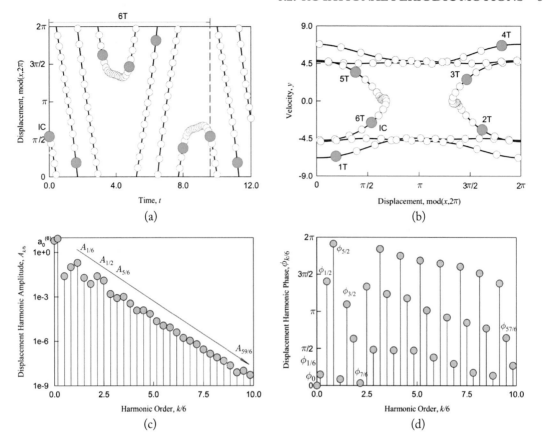

Figure 6.8: Non-travelable rotational period-6 motion with $\Omega = 3.93$: (a) displacement, (b) trajectory, (c) harmonic amplitudes, and (d) harmonic phases. ($\alpha = 4.0$, $\delta = 0.1$, $Q_0 = 5.0$). ($x_0 \approx 1.6855$, $\dot{x}_0 \approx -2.5981$).

1.0045×10^{-3}, $A_{23/6} \approx 3.5610 \times 10^{-4}$, $A_{25/6} \approx 1.2048 \times 10^{-4}$, and $A_{9/2} \approx 1.2349 \times 10^{-4}$. For the 59th harmonic term, $A_{59/6} \sim 10^{-7}$. The harmonic phases of the symmetric period-6 motion are presented in Fig. 6.8d through $\mathrm{mod}(\phi_{(2l-1)/6}, 2\pi)$ ($l = 1, 2, 3, \ldots$).

Once the afore-discussed symmetric period-6 motion encounters a saddle-node bifurcation at $\Omega \approx 3.9155$, a pair of asymmetric period-6 motions can be generated. Such asymmetric period-6 motions are non-travelable. Two paired asymmetric period-6 motions are presented in Fig. 6.9 for $\Omega = 3.91$. The initial conditions are $x_0 \approx 4.7170$, $\dot{x}_0 \approx 2.7807$ for the black branch and $x_0 \approx 1.5661$, $\dot{x}_0 \approx -2.7807$ for the red branch. The phase trajectories of the asymmetric period-6 motions for the black and red branches motions are presented in Figs. 6.9a and b, respectively. Such trajectories are skew symmetric to $(\pi, 0)$ in phase plane. From the initial

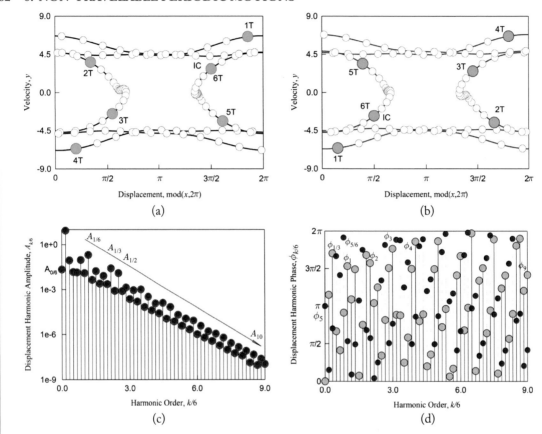

Figure 6.9: Non-travelable asymmetric period-6 motion with rotation ($\Omega = 3.91$): (a) trajectory ($x_0 \approx 4.7170$, $\dot{x}_0 \approx 2.7807$, black), (b) trajectory ($x_0 \approx 1.5661$, $\dot{x}_0 \approx -2.7807$, red), (c) harmonic amplitudes, and (d) harmonic phases. ($\alpha = 4.0$, $\delta = 0.1$, $Q_0 = 5.0$.)

condition, the asymmetric period-6 motion on the black branch goes through two full counterclockwise rotations and then two clockwise rotations to complete a complete periodic orbit. However, the asymmetric period-6 motion on the red branch motion goes through exactly opposite directions to form a closed orbit. The corresponding harmonic amplitudes and phases are illustrated in Figs. 6.9c and d, respectively. The amplitudes of all harmonic terms are the same between the two asymmetric branches except for the constant term $a_0^{(6)}$. For the black branch, the constant term $a_0^{(6)b} \approx -0.02163$ and $\mathrm{mod}(a_0^{(6)b}, 2\pi) \approx 6.26155$. On the other hand, $a_0^{(6)r} \approx 0.02163$ for the red branch. Thus, $A_{0/6}^r = A_{0/6}^b \approx 0.02163$. $A_{2l/6} = A_{l/3} \neq 0$ but $A_{(2l-1)/6} \neq 0$ ($l = 1, 2, 3, \ldots$). The main harmonic amplitudes of $A_{(2l)/6} = A_{l/3}$ are $A_{1/3} \approx 0.0869$, $A_{2/3} \approx 0.0138$, $A_1 \approx 0.0124$, $A_{4/3} \approx 4.4092 \times 10^{-3}$, $A_{5/3} \approx 2.4927 \times 10^{-3}$, $A_2 \approx 2.2882 \times 10^{-3}$, $A_{7/3} \approx 8.1905 \times 10^{-4}$, $A_{8/3} \approx 7.9190 \times 10^{-4}$, $A_3 \approx 2.3078 \times 10^{-4}$, $A_{10/3} \approx 1.5877 \times$

10^{-4}, and $A_4 \approx 4.1120 \times 10^{-5}$. The main harmonic amplitudes of $A_{(2l-1)/6}$ are $A_{1/6} \approx$ 8.5755, $A_{1/2} \approx 0.0145$, $A_{5/6} \approx 0.0924$, $A_{7/6} \approx 0.2005$, $A_{3/2} \approx 0.0168$, $A_{11/6} \approx 3.1535 \times$ 10^{-3}, $A_{13/6} \approx 0.0257$, $A_{5/2} \approx 0.0123$, $A_{17/6} \approx 1.3600 \times 10^{-3}$, $A_{19/6} \approx 9.0329 \times 10^{-4}$, $A_{7/2} \approx$ 1.0250×10^{-3}, $A_{23/6} \approx 3.3377 \times 10^{-4}$, $A_{25/6} \approx 1.2023 \times 10^{-4}$, $A_{9/2} \approx 4.1120 \times 10^{-4}$, and $A_{29/6} \approx 6.9155 \times 10^{-5}$. The corresponding harmonic phases of the two paired asymmetric motions pair are presented in Fig. 6.9d with $\phi_{k/6}^b = \mathrm{mod}(\phi_{k/6}^r + \pi, 2\pi)$ ($k = 1, 2, 3, \ldots$).

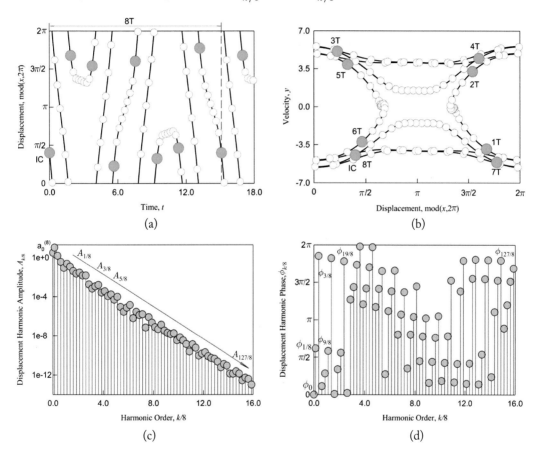

Figure 6.10: Non-travelable, rotational, symmetric period-8 motion ($\Omega = 3.33$): (a) displacement, (b) trajectory, (c) harmonic amplitudes, and (d) harmonic phases ($\alpha = 4.0$, $\delta = 0.1$, $Q_0 = 5.0$). ($t_0 = 0.0$ $x_0 \approx 1.2518$, $\dot{x}_0 \approx -4.4671$).

6.2.2 SYMMETRIC PERIOD-8 MOTIONS

In Fig. 6.10, a non-travelable symmetric rotating period-8 motion is presented for $\Omega = 3.33$. The initial condition of such motion is $x_0 \approx 1.2518$ and $\dot{x}_0 \approx -4.4671$. The displacement response of the symmetric period-8 motion is presented in Fig. 6.10a. For eight excitation periods, the symmetric periodic motion returns back the initial condition. The trajectory of such symmetric period-8 motion possesses a libration with three full clockwise rotations and then three full counter-clockwise rotations to complete a closed orbit, as shown in Fig. 6.10b. Such trajectory is symmetric to the point $(\pi, 0)$. The corresponding harmonic amplitudes of such a symmetric period-8 motion is presented in Fig. 6.10c. The constant harmonic term possesses an amplitude of $a_0^{(8)} \approx 2\pi$ and $\mathrm{mod}(a_0^{(8)}, 2\pi) \approx 0$. $A_{2l/8} = A_{l/4} = 0$ but $A_{(2l-1)/8} \neq 0$ $(l = 1, 2, \ldots)$. The main harmonic amplitudes of $A_{(2l-1)/8}$ $(l = 1, 2, \ldots)$ are $A_{1/8} \approx 10.5247$, $A_{3/8} \approx 1.5374$, $A_{5/8} \approx 0.3626$, $A_{7/8} \approx 0.0871$, $A_{9/8} \approx 0.2308$, $A_{11/8} \approx 0.1182$, $A_{13/8} \approx 0.0415$, $A_{15/8} \approx 0.0226$, $A_{17/8} \approx 0.0288$, $A_{19/8} \approx 0.0145$, $A_{21/8} \approx 0.0148$, $A_{23/8} \approx 1.7874 \times 10^{-3}$, $A_{25/8} \approx 6.7234 \times 10^{-4}$. For 127th harmonic terms, $A_{127/8} \sim 10^{-13}$. The harmonic phase distribution of $\phi_{(2l-1)/8}$ is presented in Fig. 6.10d.

CHAPTER 7

Travelable Periodic Motions

In this chapter, travelable asymmetric periodic motions on the bifurcation trees will be presented for motion complexity. There exist no any centers for such periodic motions, and the symmetric periodic motions do not exist. The travelable period-m motion has $\text{mod}(x_0, 2\pi) = \text{mod}(x_{mT}, 2\pi)$ but $x_0 \neq x_{mT}$ with $y_0 = y_{mT}$. Thus, the Fourier series of displacement cannot exist. Herein, the Fourier series of velocity will be presented to show harmonic effects on such travelable period-m motion. To demonstrate travelable periodic motions, the following formula for coordinates in the physical model are expressed by the displacement as

$$X_k = (l_0 + ka) \cos x_k, \qquad Y_k = (l_0 + ka) \sin x_k,$$
$$k = 1, 2, 3, \ldots; \qquad a = \frac{1}{1200}. \tag{7.1}$$

where l_0 is the pendulum length, ka is the fictitious function with node number k for illustration of pendulum rotation motions in the physical model. For the real physical model, we have $a = 0$. However, we cannot see the rotation complexity of displacement x. The coordinates X, Y in physical model is locations of the pendulum. Using such fictitious functions, we can easily observe the motion complexity of angular displacement, and the coefficient $a > 0$ in the fictitious function is arbitrarily chosen. Without loss of generality, for the Fourier series of velocity, the symbols for harmonic amplitudes and phases will use the same as for displacement for the non-travelable periodic motions. The periodic motions in pendulum can be characterized by the rotation and libration numbers as follows:

$$(R_+ : R_- : L), \tag{7.2}$$

where R_+ is the number of positive rotations, R_- is the number of negative rotations, and L is the number of librations. The libration number consists of positive librations and negative librations. $L = L_+ + L_-$ and $L_+ = L_-$ for periodic motions. For non-travelable period-1 motions, we have $R_+ = R_-$.

Definition 7.1 If a periodic motion in pendulum possesses R_+ positive rotations, R_- negative rotations, and L librations, then the periodic motion is called the $(R_+ : R_- : L)$-*periodic motion in pendulum.*

For a travelable period-m motion, the initial and ending points are expressed by x_0 and x_{mT}, respectively. The phase trajectory cannot form a closed orbit because of $x_0 = \text{mod}(x_{mT}, 2\pi)$

with $x_{mT} - x_0 \neq 0$. The displacement for the travelable period-m motion cannot be analyzed by the Fourier series. Thus, the velocity of for the travelable period-m motion will be analyzed by the Fourier series, and harmonic amplitudes and phases will be presented in this section.

Using similar patterns for the non-travelable periodic motions, velocities and trajectories of such travelable periodic motions will be presented numerically and analytically. The numerical and analytical results will be represented by solid curves and hollow circular symbols, respectively. The initial condition and periodic nodes are indicated by green circles. For paired asymmetric motions, the black or red color indicate the periodic motions are on the black or red branches of bifurcation trees, respectively. The acronym "IC" is initial condition. Finally, to provide a clear illustration of such travelable periodic motions, the rotation patterns of the pendulum are also presented.

7.1   TRAVELABLE PERIOD-1 TO PERIOD-4 MOTIONS

In Fig. 7.1, a pair of travelable asymmetric period-1 motions are presented for $\Omega = 4.50$. Such motions are taken from the bifurcation trees of travelable period-1 motions to chaos. The initial conditions are given by $x_0 \approx 6.0524$, $\dot{x}_0 \approx 5.7820$ and $x_0 \approx 0.2308$, $\dot{x}_0 \approx -5.7820$ for the black and red branches, respectively. The velocity responses of the two paired period-1 motions are presented in Fig. 7.1a. For a counter-clockwise, travelable period-1 motion, the velocity is always positive. For a clockwise, travelable periodic motion, the velocity is always negative. The trajectories of such paired asymmetric motions are presented in Fig. 7.1b. The two trajectories are skew symmetric to point $(\pi, 0)$ in phase plane. The initial and ending points of the period-1 motion are expressed by x_0 and x_T, respectively. For the travelable period-1 motions, the phase trajectory cannot form a closed orbit owing to $x_0 = \text{mod}(x_T, 2\pi)$ with $x_T \neq x_0$. In other words, for the paired asymmetric period-1 motions, $x_T - x_0 = 2\pi$ for the counter-clockwise rotation and $x_T - x_0 = -2\pi$ for the clockwise rotation. Based on such a periodicity condition, the displacement for such travelable periodic motion cannot be analyzed through the Fourier series. Thus, the velocity of the travelable periodic motion is used for the frequency analysis by the Fourier series. Herein, the harmonic amplitudes and phases of velocities for the paired travelable period-1 motions are presented in Figs. 7.1c,d. The constant terms of the velocity are $A_0 = a_0^b = 4.5$ and $A_0 = -a_0^r = 4.5$. Such constant terms for velocity are uniform rotation speed for the pendulum, which equals to the excitation frequency. The other harmonic terms for the two paired period-1 motions are of the same. The main harmonic amplitudes are $A_1 \approx 0.9449$, $A_2 \approx 0.3290$, and $A_3 \approx 0.0263$. The other harmonic amplitudes for velocity are $A_k \in (10^{-11}, 10^{-2})$ ($k = 4, 5, \ldots, 12$) with $A_{12} \approx 1.5 \times 10^{-10}$. The harmonic phases of velocity for the two paired period-1 motions have $\phi_k^r = \text{mod}(\phi_k^b + \pi, 2\pi)$ for $k = 0, 1, 2, \ldots$. To clearly demonstrate such travelable motions, the rotation patterns of the paired travelable period-1 motions are presented in Figs. 7.1e,f for the black and red branches, respectively. The black branch motion goes through one full counter-clockwise rotation from the initial point to complete a

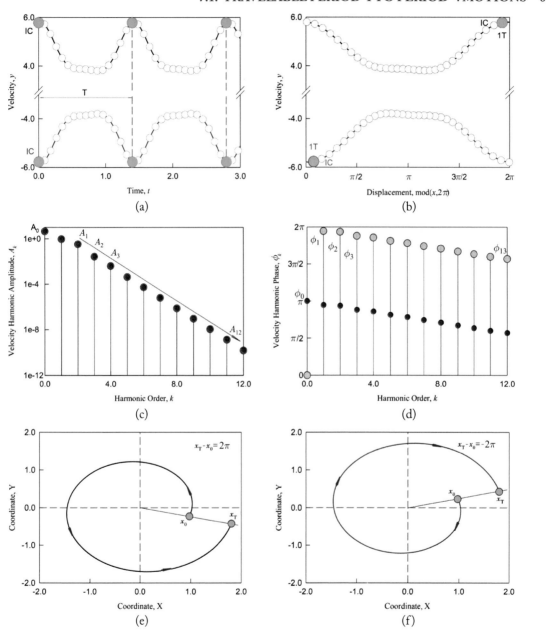

Figure 7.1: Travelable asymmetric period-1 motion ($\Omega = 4.50$): (a) velocities, (b) trajectories, (c) velocity harmonic amplitudes, (d) velocity harmonic phases, (e) pendulum rotation pattern (black), and (f) pendulum rotation pattern (red). ($\alpha = 4.0$, $\delta = 0.1$, $Q_0 = 5.0$). ($x_0 \approx 6.0524$, $\dot{x}_0 \approx 5.7820$, black) and ($x_0 \approx 0.2308$, $\dot{x}_0 \approx -5.7820$, red).

periodic motion with $x_T - x_0 = 2\pi$. On the other hand, the red branch motion goes through one full clockwise rotation to complete a periodic motion with $x_T - x_0 = -2\pi$.

The travelable asymmetric period-1 motion has a period-doubling bifurcation at $\Omega \approx 3.9640$. After the period-doubling bifurcations, two paired travelable asymmetric period-2 motions exist. Thus, two paired travelable asymmetric period-2 motions are illustrated in Fig. 7.2 for $\Omega = 3.90$. The initial conditions are $x_0 \approx 5.5361$, $\dot{x}_0 \approx 5.1493$ on the black branch and $x_0 \approx 0.7471$, $\dot{x}_0 \approx -5.1493$ on the red branch. The velocity responses of the two paired, asymmetric, travelable period-2 motions are presented in Fig. 7.2a. The trajectories of such travelable asymmetric period-2 motions are presented in Fig. 7.2b. Two phase trajectories of the paired travelable period-2 motions are skew symmetric to point $(\pi, 0)$. For the paired asymmetric period-2 motions, $x_{2T} - x_0 = 4\pi$ for the counter-clockwise rotation and $x_{2T} - x_0 = -4\pi$ for the clockwise rotation. The velocities of such travelable period-2 motions are analyzed through the Fourier series, and the harmonic amplitudes and phases of velocities for the paired travelable period-2 motions are presented in Figs. 7.2c,d. The constant terms of the velocity are $A_{0/2} = a_0^{(2)b} = 3.9$ and $A_{0/2} = -a_0^{(2)r} = 3.9$. Such constant terms for velocity are excitation frequency. The other harmonic terms for the two paired period-2 motions are of the same. The mains harmonic amplitudes are $A_{1/2} \approx 0.9052$, $A_1 \approx 1.0225$, $A_{3/2} \approx 0.2678$, $A_2 \approx 0.3530$, $A_{5/2} \approx 0.0886$, $A_3 \approx 0.0254$, $A_{7/2} \approx 0.0144$. The other harmonic amplitudes for velocity is $A_{k/2} \in (10^{-11}, 10^{-2})$ $(k = 8, 9, \ldots, 30)$ with $A_{15} \approx 4.7 \times 10^{-12}$. The harmonic phases of velocity for the two paired period-2 motion still has $\phi_{k/2}^r = \mathrm{mod}(\phi_{k/2}^b + \pi, 2\pi)$ for $k = 0, 1, 2, \ldots$. The corresponding rotation patterns are illustrated in Figs. 7.2e,f for the black and red branches, respectively. The asymmetric travelable period-2 motion on the black branch goes through two full counter-clockwise rotations to complete the periodic motion with $x_{2T} - x_0 = 4\pi$. The asymmetric travelable period-2 motion on the red branch goes through two full clockwise rotations to complete a periodic motion with $x_{2T} - x_0 = -4\pi$.

On the bifurcation tree, the travelable asymmetric period-2 motion has a period-doubling bifurcation at $\Omega \approx 3.6691$. After the period-doubling bifurcation, the paired asymmetric travelable period-4 motions are presented in Fig. 7.3 for $\Omega = 3.65$. The corresponding initial conditions are $x_0 \approx 0.9470$, $\dot{x}_0 \approx 4.9736$ for the black branch and $x_0 \approx 5.3362$, $\dot{x}_0 \approx -4.9736$ for the red branch. The velocity responses of the two paired period-4 motions are presented in Fig. 7.3a. The phase trajectories of such travelable asymmetric period-4 motion are presented in Fig. 7.3b. The two trajectories of the paired travelable period-4 motions are symmetric to point $(\pi, 0)$. For the travelable period-4 motions, the phase trajectory cannot form a closed orbit, However, $x_0 = \mathrm{mod}(x_{4T}, 2\pi)$ with $x_0 \neq x_{4T}$. For the paired asymmetric period-4 motions, $x_{4T} - x_0 = 8\pi$ for the counter-clockwise rotation and $x_{4T} - x_0 = -8\pi$ for the clockwise rotation. The harmonic amplitudes and phases of velocities for the paired travelable period-4 motions are carried out, as shown in Figs. 7.3c,d. The constant terms of the velocity are $A_{0/4} = a_0^{(4)b} = 3.65$ and $A_{0/4} = -a_0^{(4)r} = 3.65$. The mains harmonic amplitudes are $A_{1/4} \approx 0.1286$, $A_{1/2} \approx 1.8329$, $A_{3/4} \approx 0.1383$, $A_1 \approx 0.7964$, $A_{5/4} \approx 0.0315$, $A_{3/2} \approx 0.5289$,

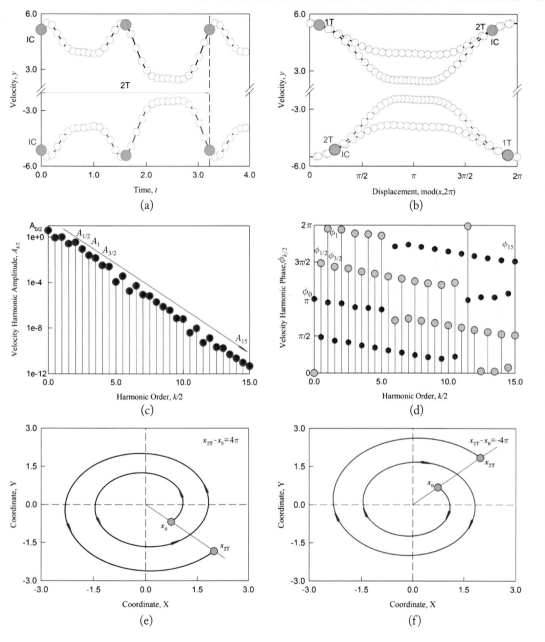

Figure 7.2: Travelable asymmetric period-2 motion ($\Omega = 3.90$): (a) velocities, (b) trajectory, (c) velocity harmonic amplitudes, (d) velocity harmonic phases, (e) pendulum rotation pattern (black), and (f) pendulum rotation pattern (red). ($\alpha = 4.0$, $\delta = 0.1$, $Q_0 = 5.0$). IC: ($x_0 \approx 5.5361$, $\dot{x}_0 \approx 5.1493$, black) and ($x_0 \approx 0.7471$, $\dot{x}_0 \approx -5.1493$, red).

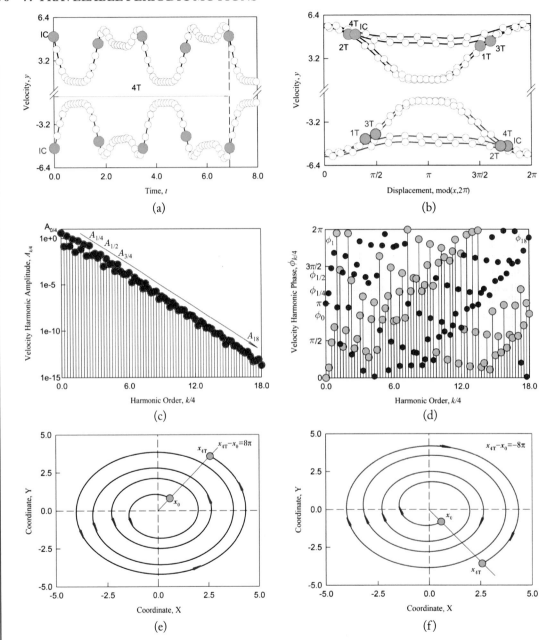

Figure 7.3: Travelable asymmetric period-4 motion ($\Omega = 3.65$): (a) velocity, (b) trajectory, (c) velocity harmonic amplitudes, (d) velocity harmonic phases, (e) pendulum rotation pattern (black), and (f) pendulum rotation pattern (red). ($\alpha = 4.0$, $\delta = 0.1$, $Q_0 = 5.0$). IC: ($x_0 \approx 0.9470$, $\dot{x}_0 \approx 4.9736$, black) and ($x_0 \approx 5.3362$, $\dot{x}_0 \approx -4.9736$, red).

$A_{7/4} \approx 0.0625$, $A_2 \approx 0.2257$, $A_{9/4} \approx 0.0123$, $A_{5/2} \approx 0.1613$, $A_{11/4} \approx 0.0140$, $A_3 \approx 0.0219$, $A_{13/4} \approx 7.7235 \times 10^{-3}$, and $A_{7/2} \approx 0.0172$. The other harmonic amplitudes for velocity is $A_{k/4} \in (10^{-14}, 10^{-2})$ $(k = 15, 16, \ldots, 72)$ with $A_{18} \approx 1.9 \times 10^{-14}$. The harmonic phases of velocity for the two paired period-4 motion still has $\phi^r_{k/4} = \mathrm{mod}(\phi^b_{k/4} + \pi, 2\pi)$ for $k = 0, 1, 2, \ldots$. The corresponding rotation patterns are then illustrated in Figs. 7.3e,f for the black and red branches, respectively. The asymmetric travelable period-4 motion on the black branch goes through four full counter-clockwise rotations to complete the periodic motion with $x_{4T} - x_0 = 8\pi$. The asymmetric travelable period-4 motion on the red branch goes through four full clockwise rotations to complete a periodic motion with $x_{4T} - x_0 = -8\pi$.

7.2 TRAVELABLE PERIOD-3 TO PERIOD-6 MOTIONS

On the bifurcation tree of asymmetric travelable period-3 motion to chaos, consider two paired asymmetric travelable period-3 motions, as presented in Fig. 7.4 for $\Omega = 4.70$. The corresponding initial conditions for the paired period-3 motions are $(x_0 \approx 1.0171, \dot{x}_0 \approx 4.8699)$ for the black branch and $(x_0 \approx 5.2661, \dot{x}_0 \approx -4.8699)$ for the red branch. The velocity responses of the two paired period-3 motions are presented in Fig. 7.4a. The phase trajectories for the two paired travelable asymmetric period-3 motions are presented in Fig. 7.4b. The two trajectories of the paired travelable period-3 motions are skew symmetric to point$(\pi, 0)$. For the paired asymmetric period-3 motions, $x_{3T} - x_0 = 6\pi$ for the counter-clockwise rotation and $x_{3T} - x_0 = -6\pi$ for the clockwise rotation. The harmonic amplitudes and phases of velocities for the travelable period-3 motions are carried out, as shown in Figs. 7.4c,d. The decay rates of the harmonic amplitudes with harmonic order for the travelable period-3 motion almost satisfy a power law. The constant terms of velocities for the paired asymmetric travelable period-3 motions are $A_{0/3} = a_0^{(3)b} = 4.7$ and $A_{0/3} = -a_0^{(3)r} = 4.7$. The main harmonic amplitudes are $A_{1/3} \approx 1.9429$, $A_{2/3} \approx 0.6282$, $A_1 \approx 0.5486$, $A_{4/3} \approx 0.4424$, $A_{5/3} \approx 0.0984$, $A_2 \approx 0.1732$, $A_{7/3} \approx 0.1478$, $A_{8/3} \approx 0.0242$, $A_3 \approx 2.3912 \times 10^{-3}$, and $A_{10/3} \approx 0.0106$. The other harmonic amplitudes for velocity is $A_{k/3} \in (10^{-14}, 10^{-2})$ $(k = 11, 12, \ldots, 48)$ with $A_{16} \approx 7.9 \times 10^{-14}$. The harmonic phases of velocity for the two paired period-3 motion have $\phi^r_{k/3} = \mathrm{mod}(\phi^b_{k/3} + \pi, 2\pi)$ for $k = 1, 2, \ldots$. The corresponding rotation patterns are then illustrated in Figs. 7.4e,f for the black and red branches, respectively. The asymmetric travelable period-3 motion on the black branch goes through three full counter-clockwise rotations to complete the periodic motion with $x_{3T} - x_0 = 6\pi$. The asymmetric travelable period-3 motion on the red branch goes through three full clockwise rotations to complete a periodic motion with $x_{3T} - x_0 = -6\pi$.

On the bifurcation tree of asymmetric travelable period-3 motion to chaos, such a period-3 motion has a period-doubling bifurcation at $\Omega \approx 4.6675$. After the period-doubling bifurcation, two paired asymmetric period-6 motions are generated. Such travelable asymmetric period-6 motions are illustrated in Fig. 7.5 for $\Omega = 4.65$. The initial conditions are given by $x_0 \approx 1.1440$, $\dot{x}_0 \approx 4.7474$ for the black branch and $x_0 \approx 5.1392$, $\dot{x}_0 \approx -4.7474$ for the red branch. The velocity responses of the two paired period-6 motions are presented in Fig. 7.5a.

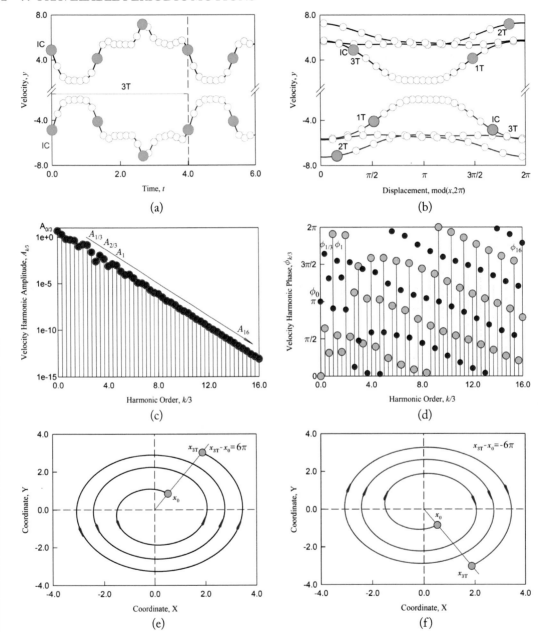

Figure 7.4: Travelable asymmetric period-3 motion ($\Omega = 4.70$): (a) velocities, (b) trajectories, (c) velocity harmonic amplitudes, (d) velocity harmonic phases, (e) pendulum rotation pattern (black), and (f) pendulum rotation pattern (red). ($\alpha = 4.0$, $\delta = 0.1$, $Q_0 = 5.0$). IC: ($x_0 \approx 1.0171$, $\dot{x}_0 \approx 4.8699$, black) and ($x_0 \approx 5.2661$, $\dot{x}_0 \approx -4.8699$, red).

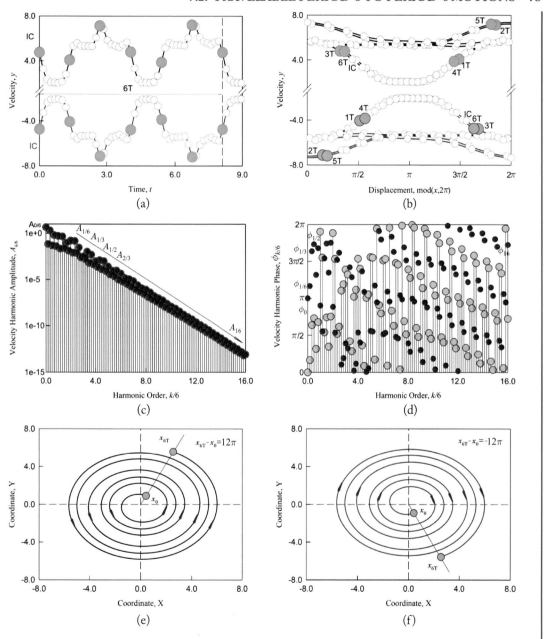

Figure 7.5: Travelable asymmetric period-6 motion ($\Omega = 4.65$): (a) velocities, (b) trajectories, (c) velocity harmonic amplitudes, (d) velocity harmonic phases, (e) pendulum rotation pattern (black), and (f) pendulum rotation pattern (red). ($\alpha = 4.0$, $\delta = 0.1$, $Q_0 = 5.0$). IC: ($x_0 \approx 1.1440$, $\dot{x}_0 \approx 4.7474$, black) and ($x_0 \approx 5.1392$, $\dot{x}_0 \approx -4.7474$, red).

The phase trajectories for the two paired travelable asymmetric period-6 motions are presented in Fig. 7.2b. The two trajectories of the paired travelable period-6 motions are skew symmetric to point$(\pi, 0)$. For the travelable period-6 motions, the phase trajectory still cannot form a closed orbit. For the paired asymmetric period-6 motions, $x_{6T} - x_0 = 12\pi$ for the counter-clockwise rotation and $x_{6T} - x_0 = -12\pi$ for the clockwise rotation. The harmonic amplitudes and phases of velocities for the travelable period-6 motions are carried out, as shown in Figs. 7.5c,d. The constant terms of velocities for the paired asymmetric travelable period-6 motions are $A_{0/6} = a_0^{(6)b} = 4.65$ and $A_{0/6} = -a_0^{(6)r} = 4.65$. The main harmonic amplitudes are $A_{1/6} \approx 0.0700$, $A_{1/3} \approx 2.0200$, $A_{1/2} \approx 0.0491$, $A_{2/3} \approx 0.6432$, $A_{5/6} \approx 0.0400$, $A_1 \approx 0.5233$, $A_{7/6} \approx 0.0219$, $A_{4/3} \approx 0.4602$, $A_{3/2} \approx 0.0243$, $A_{5/3} \approx 0.0959$. $A_{11/3} \approx 0.0133$, $A_2 \approx 0.1646$, $A_{13/6} \approx 7.839010 \times 10^{-3}$, $A_{7/3} \approx 0.15217$, $A_{5/2} \approx 7.3754 \times 10^{-3}$, $A_{8/3} \approx 0.0290$, $A_{17/6} \approx 2.4656 \times 10^{-3}$, $A_3 \approx 3.0799 \times 10^{-3}$, $A_{19/6} \approx 1.6441 \times 10^{-3}$, and $A_{10/3} \approx 0.0103$. The other harmonic amplitudes for velocity is $A_{k/6} \in (10^{-14}, 10^{-2})$ $(k = 1, 2, \ldots, 96)$ with $A_{16} \approx 8.1 \times 10^{-14}$. The harmonic phases of velocity for the two paired period-6 motion are $\phi_{k/6}^r = \mathrm{mod}(\phi_{k/6}^b + \pi, 2\pi)$ for $k = 0, 1, 2, \ldots$. The corresponding rotation patterns are then illustrated in Figs. 7.5e,f for the black and red branches, respectively. The asymmetric travelable period-6 motion on the black branch goes through six full counter-clockwise rotations to complete the periodic motion with $x_{6T} - x_0 = 12\pi$. The asymmetric travelable period-6 motion on the red branch goes through six full clockwise rotations to complete a periodic motion with $x_{6T} - x_0 = -12\pi$.

7.3 TRAVELABLE PERIOD-5 MOTIONS

For complexity illustration of periodic motions, the paired asymmetric travelable period-5 motions are illustrated in Fig. 7.6 for $\Omega = 3.95$. The corresponding initial conditions are $x_0 \approx 4.8508$, $\dot{x}_0 \approx 4.3772$ for the black branch and $x_0 \approx 1.4323$, $\dot{x}_0 \approx -4.3772$ for the red branch. The velocity responses of the two paired period-5 motions are also presented first in Fig. 7.3a. The trajectories of the two travelable asymmetric period-5 motions are presented in Fig. 7.6b. The two trajectories of the paired travelable period-5 motions are skew symmetric to point $(\pi, 0)$. The initial and ending points are indicated by x_0 and x_{5T}, respectively. The harmonic amplitudes and phases of velocities for the travelable period-3 motions are carried out, as shown in Figs. 7.6c,d. The constant terms of velocities for the paired asymmetric travelable period-5 motions are $A_{0/5} = a_0^{(5)b} = 3.95$ and $A_{0/5} = -a_0^{(5)r} = 3.95$. The mains harmonic amplitudes are $A_{1/5} \approx 0.0990$, $A_{2/5} \approx 1.1413$, $A_{3/5} \approx 0.7809$, $A_{4/5} \approx 0.1639$, $A_1 \approx 0.8740$, $A_{6/5} \approx 0.0390$, $A_{7/5} \approx 0.2805$, $A_{8/5} \approx 0.2533$, $A_{9/5} \approx 0.0928$, $A_2 \approx 0.2823$, $A_{11/5} \approx 0.0104$, $A_{12/5} \approx 0.1026$, $A_{13/5} \approx 0.0606$, $A_{14/5} \approx 0.0313$, and $A_3 \approx 6.004 \times 10^{-3}$. The other harmonic amplitudes for velocity is $A_{k/5} \in (10^{-13}, 10^{-3})$ $(k = 15, 16, \ldots, 80)$ with $A_{16} \approx 5.88 \times 10^{-13}$. The harmonic phases of velocity for the two paired period-5 motion are $\phi_{k/5}^r = \mathrm{mod}(\phi_{k/5}^b + \pi, 2\pi)$ for $k = 0, 1, 2, \ldots$. The corresponding rotation patterns are then illustrated in Figs. 7.6e,f for the

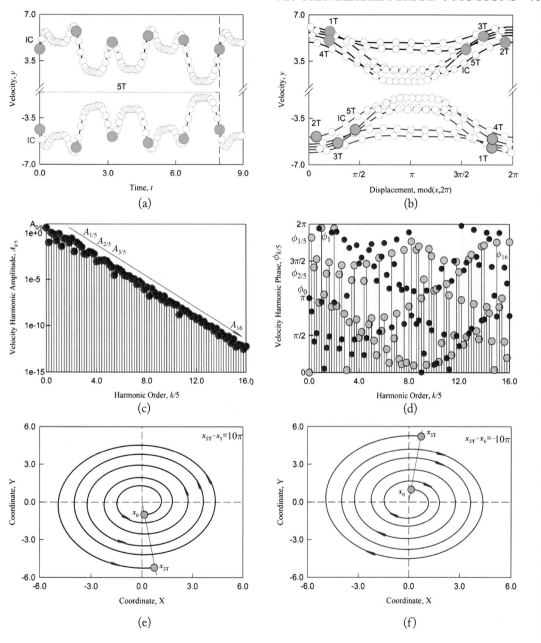

Figure 7.6: Travelable asymmetric period-5 motion ($\Omega = 3.95$, $t_0 = 0.0$): (a) velocities, (b) trajectories, (c) harmonic amplitudes, (d) velocity harmonic phases, (e) pendulum rotation pattern (black), and (f) pendulum rotation pattern (red). ($\alpha = 4.0$, $\delta = 0.1$, $Q_0 = 5.0$). ($x_0 \approx 4.8508$, $\dot{x}_0 \approx 4.3772$, black) and ($x_0 \approx 1.4324$, $\dot{x}_0 \approx -4.3772$, red).

black and red branches, respectively. The asymmetric travelable period-5 motion on the black branch goes through five full counter-clockwise rotations to complete the periodic motion with $x_{5T} - x_0 = 10\pi$. The asymmetric travelable period-5 motion on the red branch goes through five full clockwise rotations to complete a periodic motion with $x_{5T} - x_0 = -10\pi$.

References

[1] Lagrange, J. L. (1788). *Mecanique Analytique*, vol. 2 (edition Albert Balnchard: Paris, 1965). DOI: 10.1017/cbo9780511701788. 1

[2] Poincare, H. (1899). *Methodes Nouvelles de la Mecanique Celeste*, vol. 3. Gauthier-Villars: Paris. DOI: 10.1007/bf02742713. 1

[3] van der Pol, B. (1920). A theory of the amplitude of free and forced triode vibrations. *Radio Review*, 1:701–710, 754–762. 1

[4] Fatou, P. (1928). Sur le mouvement d'un systeme soumis à des forces a courte periode. *Bulletin de la Société Mathématique*, 56:98–139. DOI: 10.24033/bsmf.1131. 1

[5] Krylov, N. M. and Bogolyubov, N. N. (1935). Methodes approchees de la mecanique non-lineaire dans leurs application a l'Aeetude de la perturbation des mouvements periodiques de divers phenomenes de resonance s'y rapportant. *Academie des Sciences d'Ukraine: Kieve*. (in French). 1

[6] Hayashi, C. (1964). *Nonlinear Oscillations in Physical Systems*. McGraw-Hill Book Company: New York. DOI: 10.1515/9781400852871. 1

[7] Barkham, P. G. D. and Soudack, A. C. (1969). An extension to the method of Krylov and Bogoliubov. *International Journal of Control*, 10:377–392. DOI: 10.1080/00207176908905841. 1

[8] Garcia-Margallo, J. and Bejarano, J. D. (1987). A generalization of the method of harmonic balance. *Journal of Sound and Vibration*, 116:591–595. DOI: 10.1016/s0022-460x(87)81390-1. 1

[9] Yuste, S. B. and Bejarano, J. D. (1986). Construction of approximate analytical solutions to a new class of non-linear oscillator equations. *Journal of Sound and Vibration*, 110(2):347–350. DOI: 10.1016/s0022-460x(86)80215-2. 1

[10] Yuste, S. B. and Bejarano, J. D. (1990). Improvement of a Krylov–Bogoliubov method that uses Jacobi elliptic functions. *Journal of Sound and Vibration*, 139(1):151–163. DOI: 10.1016/0022-460x(90)90781-t. 1

[11] Coppola, V. T. and Rand, R. H. (1990). Averaging using elliptic functions: Approximation of limit cycle. *Acta Mechanica*, 81:125–142. DOI: 10.1007/bf01176982. 1

[12] Leven, R. W. and Koch, B. P. (1981). Chaotic behavior of a parametrically excited damped pendulum. *Physics Letters*, 86A(2):71–74. DOI: 10.1016/0375-9601(81)90167-5. 2

[13] McLaughlin, J. B. (1981). Period-doubling bifurcations and chaotic motion for a parametrically forced pendulum. *Journal of Statistical Physics*, 24(2):375–388. DOI: 10.1007/bf01013307. 2

[14] Koch, B. P. and Leven, R. W. (1985). Subharmonic and homoclinic bifurcations in a parametrically forced pendulum. *Physica*, 16D:1–13. DOI: 10.1016/0167-2789(85)90082-x. 2

[15] Clifford, M. J. and Bishop, S. R. (1995). Rotating periodic orbits of the parametrically excited pendulum. *Physics Letters A*, 201:191–196. DOI: 10.1016/0375-9601(95)00255-2. 2

[16] Clifford, M. J. and Bishop, S. R. (1996). Locating oscillatory orbits of the parametrically-excited pendulum. *Journal of Australian Mathematics Society Series B*, 37:309–319. DOI: 10.1017/s0334270000010687. 2

[17] Sanjuan, M. A. F. (1998). Subharmonic bifurcations in a pendulum parametrically excited by a non-harmonic perturbation. *Chao, Solitons and Fractals*, 9(6):995–1003. DOI: 10.1016/s0960-0779(97)00181-1. 2

[18] Luo, A. C. J. (2000). Chaotic motions in the resonant separatrix band of a parametrically excited pendulum. *Communications in Nonlinear Science and Numerical Simulation*, 5(4):135–140. DOI: 10.1016/s1007-5704(00)90024-8. 2

[19] Luo, A. C. J. (2001). Resonance and stochastic layer in a parametrically excited pendulum. *Nonlinear Dynamics*, 26:355–367. DOI: 10.1023/A:1012996229150. 2

[20] Garira, W. and Bishop, S. R. (2003). Rotating solutions of the parametrically excited pendulum. *Journal of Sound and Vibration*, 263:233–239. DOI: 10.1016/s0022-460x(02)01435-9. 2

[21] Xu, X., Wiercigroch, M., and Cartmell, M. P. (2005). Rotating orbits of a parametrically-excited pendulum. *Chaos, Solitons and Fractals*, 23:1537–1548. DOI: 10.1016/j.chaos.2004.06.053. 2

[22] Xu, X. and Weircigroch, M. (2007). Approximate analytical solutions for oscillatory and rotational motion of a parametric pendulum. *Nonlinear Dynanmics*, 47:311–320. DOI: 10.1007/s11071-006-9074-4.

[23] Lenci, S., Pavlovskaia, E., Rega, G., and Wiercigroch, M. (2008). Rotating solutions and stability of parametric pendulum by perturbation method. *Journal of Sound and Vibration*, 310:243–259. DOI: 10.1016/j.jsv.2007.07.069. 2

[24] Lu, C. Q. (2007). Chaos of a parametrically excited undamped pendulum. *Communications in Nonlinear Science and Numerical Simulation*, 12:45–57. DOI: 10.1016/j.cnsns.2006.01.004. 2

[25] Luo, A. C. J. (2012). *Continuous Dynamical Systems*. HEP/L&H Scientific: Beijing/Glen Carbon. 2

[26] Luo, A. C. J. (2014). On analytical routes to chaos in nonlinear systems. *International Journal of Bifurcation and Chaos*, 24(4), no. 1430013, 28 pages. DOI: 10.1142/s0218127414300134. 2

[27] Luo, A. C. J. (2014). *Toward Analytical Chaos in Nonlinear Dynamical Systems*. Wiley. DOI: 10.1002/9781118887158.

[28] Luo, A. C. J. (2014). *Analytical Routes to Chaos in Nonlinear Engineering*. Wiley. DOI: 10.1002/9781118883938. 2

[29] Luo, A. C. J. and Huang, J. (2012). Approximate solutions of periodic motions in nonlinear systems via a generalized harmonic balance. *Journal of Vibration and Control*, 18:1661–1671. DOI: 10.1177/1077546311421053. 2

[30] Luo, A. C. J. and Huang, J. Z. (2012b). Analytical dynamics of period-*m* flows and chaos in nonlinear systems. *International Journal of Bifurcation and Chaos*, 22(4), no. 1250093, 29 pages. DOI: 10.1142/s0218127412500939. 2

[31] Luo, A. C. J. and Huang, J. Z. (2012). Analytical routes of period-1 motions to chaos in a periodically forced Duffing oscillator with a twin-well potential. *Journal of Applied Nonlinear Dynamics*, 1:73–108. DOI: 10.5890/jand.2012.02.002. 2

[32] Luo, A. C. J. and Huang, J. Z. (2012). Unstable and stable period-m motions in a twin-well potential Duffing oscillator. *Discontinuity, Nonlinearity and Complexity*, 1:113–145. DOI: 10.5890/dnc.2012.03.001.

[33] Luo, A. C. J. and Huang, J. Z. (2013). Analytical solutions for asymmetric periodic motions to chaos in a hardening Duffing oscillator. *Nonlinear Dynamics*, 72:417–438. DOI: 10.1007/s11071-012-0725-3.

[34] Luo, A. C. J. and Huang, J. Z. (2013). Analytical period-3 motions to chaos in a hardening Duffing oscillator. *Nonlinear Dynamics*, 73:1905–1932. DOI: 10.1007/s11071-013-0913-9.

[35] Luo, A. C. J. and Huang, J. Z. (2013). Asymmetric periodic motions with chaos in a softening Duffing oscillator. *International Journal of Bifurcation and Chaos*, 23(5), no: 1350086, 31 pages. DOI: 10.1115/imece2017-70824.

[36] Luo, A. C. J. and Huang, J. Z. (2014). Period-3 motions to chaos in a softening Duffing oscillator. *International Journal of Bifurcation and Chaos*, 24, no. 1430010, 26 pages. DOI: 10.1142/s0218127414300109. 2, 17

[37] Luo, A. C. J. and Yu, B. (2013). Complex period-1 motions in a periodically forced, quadratic nonlinear oscillator. *Journal of Vibration of Control*, 21(5):907–918. DOI: 10.1177/1077546313490525. 2

[38] Luo, A. C. J. and Laken, A. B. (2013). Analytical solutions for period-m motions in a periodically forced van der Pol oscillator. *International Journal of Dynamics and Control*, 1(2):99–155. DOI: 10.1115/imece2012-86589. 2

[39] Luo, A. C. J. and Laken, A. B. (2014). An approximate solution for period-1 motions in a periodically forced van der Pol oscillator. *ASME Journal of Computational and Nonlinear Dynamics*, 9(3), no. 031001, 7 pages. DOI: 10.1115/1.4026425. 2

[40] Luo, A. C. J. and Laken, A. B. (2014). Period-m motions and bifurcation in a periodically forced van der Pol-Duffing oscillator. *International Journal of Dynamics and Control*, 2:474–493. DOI: 10.1007/s40435-014-0058-9. 2

[41] Luo, A. C. J. and Yu, B. (2015). Bifurcation trees of period-1 motions to chaos in a two-degree-of-freedom, nonlinear oscillator. *International Journal of Bifurcation and Chaos*, 25(13), no: 1550179, 40 pages. DOI: 10.1142/s0218127415501795. 2

[42] Yu, B. and Luo, A. C. J. (2017). Analytical period-1 motions to chaos in a two-degree-of-freedom oscillator with a hardening nonlinear spring, *International Journal of Dynamics and Control*, 5(3):436–453. DOI: 10.1007/s40435-015-0216-8. 2

[43] Luo, A. C. J. (2013). Analytical solutions for periodic motions to chaos in nonlinear systems with/without time-delay. *International Journal of Dynamics and Control*, 1:330–359. DOI: 10.1007/s40435-013-0024-y. 2

[44] Luo, A. C. J. and Jin, H. X. (2014). Bifurcation trees of period-m motion to chaos in a Time-delayed, quadratic nonlinear oscillator under a periodic excitation. *Discontinuity, Nonlinearity, and Complexity*, 3:87–107. DOI: 10.5890/dnc.2014.03.007. 2

[45] Luo, A. C. J. and Jin, H. X. (2014). Period-m motions to chaos in a periodically forced Duffing oscillator with a time-delay feedback. *International Journal of Bifurcation and Chaos*, 24(10), no: 1450126, 20 pages. DOI: 10.1142/s0218127414501260. 2

[46] Luo, A. C. J. and Jin, H. X. (2015). Complex period-1 motions of a periodically forced Duffing oscillator with a time-delay feedback. *International Journal of Dynamics and Control*, 3:325–340. DOI: 10.1007/s40435-014-0091-8.

[47] Luo, A. C. J. and Jin, H. X. (2015). Period-3 motions to chaos in a periodically forced Duffing oscillator with a linear time-delay. *International Journal of Dynamics and Control*, 3:371–388. DOI: 10.1007/s40435-014-0116-3. 2

[48] Luo, A. C. J. (2015). Periodic flows to chaos based on discrete implicit mappings of continuous nonlinear systems. *International Journal of Bifurcation and Chaos*, 25(3), no. 1550044, 62 pages. DOI: 10.1142/s0218127415500443. 2, 5, 9, 16

[49] Luo, A. C. J. and Guo, Y. (2015). A semi-analytical prediction of periodic motions in duffing oscillator through mapping structures. *Discontinuity, Nonlinearity, and Complexity*, 4(2):13–44. DOI: 10.5890/dnc.2015.06.002. 2

[50] Guo, Y. and Luo, A. C. J. (2015). On complex periodic motions and bifurcations in a periodically forced, damped, hardening duffing oscillator. *Chaos, Solitons and Fractals*, 81:378–399. DOI: 10.1016/j.chaos.2015.10.004. 2

[51] Luo, A. C. J. and Xing, S. Y. (2016). Symmetric and asymmetric period-1motion in a periodically forced, time-delayed, hardening Duffing oscillator. *Nonlinear Dynamics*, 85(2):1141–1161. DOI: 10.1007/s11071-016-2750-0. 3

[52] Luo, A. C. J. and Xing, S. Y. (2016). Multiple bifurcation trees of period-1 motions to chaos in a periodically forced, time-delayed, hardening Duffing oscillator. *Chaos, Solitons and Fractals*, 89:405–434. DOI: 10.1016/j.chaos.2016.02.005. 3

[53] Luo, A. C. J. and Xing, S. Y. (2017). Time-delay effects on periodic motions in a Duffing oscillator. *Chaotic, Fractional, and Complex Dynamics: New Insights and Perspectives*, by Mark Edelman, Elbert E. N. Macau, and Miguel A. F. Sanjuan, Eds. Springer. DOI: 10.1007/978-3-319-68109-2_5. 3

[54] Xing, S. Y. and Luo, A. C. J. (2017). Towards infinite bifurcation trees of period-1 motions to chaos in a time-delayed, twin-well duffing oscillator. *Journal of Vibration Testing and System Dynamics*, 1(4):353–392. DOI: 10.5890/jvtsd.2017.12.006. 3

[55] Xing, S. Y. and Luo, A. C. J. (2018). On possible infinite bifurcation trees of period-3 motions to chaos in a time-delayed, twin-well Duffing oscillator. *International Journal of Dynamics and Control*, 6(4):1429–1464. DOI: 10.1007/s40435-018-0418-y. 3

[56] Xu, Y. Y. and Luo, A. C. J. (2019). A series of symmetric period-1 motions to chaos in a two-degree-of-freedom van der Pol-duffing oscillator. *Journal of Vibration Testing and System Dynamics*, 2(2):119–153. DOI: 10.5890/jvtsd.2018.06.003. 3

[57] Xu, Y. Y. and Luo, A. C. J. (2019). Sequent period-$(2m - 1)$ motions to chaos in the van del Pol oscillator. *International Journal of Dynamics and Control*, 7(3):795–807. DOI: 10.1007/s40435-018-0468-1. 3

[58] Guo, Y. and Luo, A. C. J. (2016). Routes of periodic motions to chaos in a periodically forced pendulum. *International Journal of Dynamics and Control*, 5(3):551–569. DOI: 10.1007/s40435-016-0249-7. 3

[59] Luo, A. C. J. and Guo, Y. (2016). Periodic motions to chaos in pendulum. *International Journal of Bifurcation and Chaos*, 26(9), no. 1650159, 64 pages. DOI: 10.1142/s0218127416501595. 3

[60] Guo, Y. and Luo, A. C. J. (2017). Complete bifurcation trees of a parametrically driven pendulum. *Journal of Vibration Testing and System Dynamics*, 1(2):93–134. DOI: 10.5890/jvtsd.2017.06.001. 3

Authors' Biographies

YU GUO

Dr. Yu Guo was born on January 14, 1984. He is currently working at Midwestern State University Texas as an Associate Professor. He previously worked at Caterpillar Inc. as an engine structural and dynamics engineer. Dr. Guo was originally from China. He has his B.S. in Computer Engineering from South China University of Technology. He received his M.S. and Ph.D. in Mechanical Engineering at Southern Illinois University Edwardsville. Dr. Guo conducts research focusing on nonlinear vibration and impact dynamics. He has published 14 peer-reviewed journal papers, more than 15 conference articles, 4 book chapters, and 1 monograph. He has also conducted many professional presentations or invited lectures all over the world.

ALBERT C.J. LUO

Professor Albert C.J. Luo has worked at Southern Illinois University Edwardsville. For over 30 years, Dr. Luo's contributions on nonlinear dynamical systems and mechanics lie in: (i) the local singularity theory for discontinuous dynamical systems; (ii) dynamical systems synchronization; (iii) analytical solutions of periodic and chaotic motions in nonlinear dynamical systems; (iv) the theory for stochastic and resonant layer in nonlinear Hamiltonian systems; and (v) the full nonlinear theory for a deformable body. Such contributions have been scattered into 20 monographs and over 300 peer-reviewed journal and conference papers. Dr. Luo has served as an editor for the journal *Communications in Nonlinear Science and Numerical Simulation*, and book series on Nonlinear Physical Science (HEP) and Nonlinear Systems and Complexity (Springer). Dr. Luo was an editorial member for *IMeCh E Part K Journal of Multibody Dynamics* and *Journal of Vibration and Control*; and has also organized over 30 international symposiums and conferences on dynamics and control.

Printed in the United States
by Baker & Taylor Publisher Services